Lecture Notes in Mathematics 1696

Springer
Berlin
Heidelberg
New York
Barcelona
Budapest
Hong Kong
London
Milan
Paris
Singapore
Tokyo

Elisabeth Bouscaren (Ed.)

Model Theory and Algebraic Geometry

An introduction to E. Hrushovski's proof of the geometric Mordell-Lang conjecture

Editor

Elisabeth Bouscaren
Université Paris 7 – C.N.R.S.
UFR de Mathématiques
2 Place Jussieu, case 7012
F-75251 Paris Cedex 05, France
e-mail: elibou@logique.jussieu.fr

Cataloging-in Publication Data available.

Corrected 2nd Printing 1999

Mathematics Subject Classification (1991):
11U09, 03C60, 11G10, 14G99, 03C45

ISSN 0075-8434
ISBN 3-540-64863-1 Springer-Verlag Berlin Heidelberg New York

Typesetting: Camera-ready TEX output by the author
SPIN: 10735063 41/3143-543210 - Printed on acid-free paper

Preface

Introduction

Model theorists have often joked in recent years that the part of mathematical logic known as "pure model theory" (or stability theory), as opposed to the older and more traditional "model theory applied to algebra", turns out to have more and more to do with other subjects of mathematics and to yield genuine applications to combinatorial geometry, differential algebra and algebraic geometry.

We illustrate this by presenting the very striking application to diophantine geometry due to Ehud Hrushovski: using model theory, he has given the first proof valid in all characteristics of the "Mordell-Lang conjecture for function fields" (*The Mordell-Lang conjecture for function fields*, Journal AMS 9 (1996), 667-690). More recently he has also given a new (model theoretic) proof of the Manin-Mumford conjecture for semi-abelian varieties over a number field. His proof yields the first effective bound for the cardinality of the finite sets involved (*The Manin-Mumford conjecture*, preprint).

There have been previous instances of applications of model theory to algebra or number theory, but these applications had in common the feature that their proofs used a lot of algebra (or number theory) but only very basic tools and results from the model theory side: compactness, first-order definability, elementary equivalence...

Hrushovski's results are not only interesting as such but also due to the nature of their proofs which use in an essential way most of the beautiful and sophisticated recent developments of model theory. In fact he shows that these questions of diophantine geometry can be naturally integrated into the abstract framework which has been developed in model theory these last years.

Let us go back a few years to recall a bit of informal history, without any attempt at exhaustive coverage or precise attribution. Model theory during its first years of existence, from the 1930's to the 60's, was traditionaly related to universal algebra. Indeed, in model theory one considers classes of "abstract structures", of which the classical mathematical structures will be particular instances. These structures come equipped with a distinguished class of subsets, the definable sets, which as their name indicates, are "defined" from the basic operations of the structure.

Then in the sixties, first with M. Morley and then with the colossal work of S. Shelah, a new perspective arose around two main lines: the idea was to classify structures according to the type of combinatorial objects one could define in them (infinite orderings, infinite trees...) and also to use this classification to assign dimensions to the definable sets in certain cases. This was the birth of "classification theory" or "stability theory" , considered as "pure model theory", in contrast with the Robinson type of applied model theory. The inspiration here was primarily of a set theoretic and infinite combinatorial nature.

A second change of perspective took place in the 80's with B. Zilber in particular and then since 1986 with the work of Hrushovski. This was the birth of "geometric stability" or "geometric model theory", where the inspiration comes from combinatorial geometries and algebraic geometry. While integrating stability theory in the style of Shelah, one focuses here on the study of definable sets with finite dimension and their classification according to which type of algebraic objects (groups, fields) or geometries can be defined in them.

The following trichotomy turns out to be very relevant: structures where no group is definable, groups of linear type (which behave like vector spaces) and field-like structures. One context in which this trichotomy is particularly meaningful comes out of the beautiful work of Hrushovski and Zilber on "Zariski geometries", where they characterize abstractly amongst noetherian topologies the ones arising from the Zariski topology of an algebraic curve (over an algebraically closed field).

As we will see in this volume, the Mordell-Lang conjecture really says that certain subgroups of a semi-abelian variety are of linear type. Hrushovski first sets the question in the adequate framework, differentially closed fields in characteristic zero and (non algebraically closed) separably closed fields in characteristic p. He proceeds to apply the powerful tools that have been developed around this trichotomy in the past years. One of the interesting features of this proof is the fact that it is uniform for both characteristic zero and characteristic p, only the basic settings are different. It should also be noted that in fact the characteristic zero case can be deduced from the characteristic p case, as is shown by Hrushovski in the last chapter of this volume.

Suggestions for further reading

In this volume we focus on one result, the Mordell-Lang conjecture for function fields and present only those parts of geometric model theory which are relevant and useful for this proof. In particular we say nothing of the more recent work of Hrushovski on the Manin-Mumford conjecture which also fits in this general model-theoretic framework (the adequate setting in this case is algebraically closed fields with an automorphism, ACFA). The reader can find a partial exposition of these results in *ACFA and the Manin-Mumford Conjecture*, A. Pillay, in Algebraic Model Theory, B. Hart, A. Lachlan and M. Valeriote ed., NATO ASI Series C 496, Kluwer 1997, pp: 195-205.

For a recent survey on model theory and diophantine geometry with some sketches of proof (in particular for the characteristic p case of Mordell-Lang), see *Model theory and diophantine geometry*, A. Pillay, Bull. Am. Math. Soc. 34 (1997), pp: 405-422.

There is another related but different line of work which has seen substantial development recently, namely the relationship between differential algebra (with work in particular of P. Cassidy and of A. Buium) and the model theoretic point of view on differentially closed fields. This subject is not treated in this volume except when it relates to our own purpose (in the chapters of C. Wood and D. Marker). For more on this subject, see for example :

- *Model Theory of Fields*, D. Marker, M. Messmer and A. Pillay, Lecture Notes in Logic 5, Springer, 1996.
- *Model Theory, Differential Algebra and Number Theory*, A. Pillay, in Proceedings of ICM 94 Zurich, Birkhauser 1996.

Presentation of the volume

A few months after Hrushovski announced his model theoretic proof of the Mordell-Lang conjecture, it became apparent that there were some natural obstacles to the understanding of these new exciting results. In order to understand one needs a minimal knowledge of the basics of algebraic geometry but, more importantly, a good knowledge of the recent developments of model theory, or more precisely, of geometric stability theory.

Thus the idea arose naturally to present a more or less self-contained exposition of this proof. This was concretized first as a series of coordinated lectures in a summer school organized in September 1994 in Manchester (UK) in the framework of "RESMOD" and devoted exclusively to this subject (RESMOD is the acronym for the European Human Capital and Mobility Network on "Model Theory and its applications" coordinated by the Équipe de Logique Mathématique CNRS-Université Paris 7 (1994-1997)). This workshop was intended mainly for young researchers in model theory, and only a very basic knowledge of classical model theory was assumed. After the success of this workshop and the interest shown in the community, the need became apparent for something more elaborate than the notes that were distributed at the time of the workshop. Each of the speakers at the workshop agreed to write a chapter of this book on the subject of his/her lectures, and to collaborate with the others in order to obtain a progressive and coherent presentation. Hrushovski himself, who had not participated to the Summer School, contributed a final chapter, where he shows that "characteristic p implies characteristic 0".

The aim of this volume is to take a mathematician with a very basic knowledge of both model theory and algebraic geometry and to introduce her/him to the relationship between geometric stability theory and algebraic geometry, finishing with the detailed exposition of Hrushovski's proof of the Mordell-Lang conjecture for function fields.

In the hope that this might be used also by mathematicians with no previous knowledge whatsoever of model theory, a first chapter gives an informal presentation of the main basic definitions and results of "classical" model theory.

In order to be really self-contained and to avoid certain technical difficulties that might hinder the reader's understanding, we have chosen in this volume to make two restrictions. First, we restrict ourselves to the case of abelian varieties, but Hrushovski's proof really works in the same way for semi-abelian varieties with just a little extra effort needed on the model theoretic side in the characteristic p case. Secondly, we have chosen to present exhaustively the characteristic zero case, which is a little easier to describe from the model theoretic point of view. The characteristic p case goes through the same steps. We give here the details of the characteristic p setting (separably closed fields of finite invariant) and explain at the end where the main differences lie. We believe that a reader who understands both the characteristic zero case and the characteristic p setting should be able to see the characteristic p case quite well.

Some of the chapters are totally self-contained, with complete proofs. Others are surveys, either because the subject is too vast or because there already exist some good (and accessible) references. In the case of surveys, special effort has been made so that all results needed further along in the book are clearly stated and some comments and examples are given in order to give the reader some intuitive understanding of the subject. Each chapter has its own reference list.

Description of the chapters

- ***Introduction to model theory:*** an informal introduction to the very basic concepts of model theory, to help the reader with no previous knowledge of the subject.

- ***Introduction to stability theory and Morley rank:*** a self-contained detailed presentation, with all proofs included, of the necessary classical material from stability. The presentation and the choice of material are adapted to the context of this book.

- ***Omega-stable groups:*** classical results on groups of finite Morley rank as well as more recent results on one-based groups. Again this is self-contained, referring only to the previous chapter. Complete proofs are given.

- ***Model theory of algebraically closed fields:*** a survey, with proofs included, of the model theoretic approach to algebraically closed fields, algebraic varieties and algebraic groups. This is intended both as an introduction to the model theoretic point of view on basic algebraic geometry for the geometers and as an introduction to what is needed of algebraic geometry for the model theorists.

- ***Introduction to abelian varieties and the Mordell-Lang conjecture:*** an introduction to the conjecture and related questions, aimed at non specialists of algebraic geometry, with a survey of the main classical properties of abelian varieties which are needed in this volume. This chapter presents no proofs but includes a detailed bibliography of the subject, with comments.

- ***The model-theoretic content of Lang's conjecture:*** this short chapter explains (with proof) what the model theoretic equivalent of the Mordell-Lang conjecture is.

- ***Zariski geometries:*** a survey of the important paper by Hrushovski and Zilber. It would be impossible to give proofs here so the stress is on giving examples and comments that can shed some light on the results. At the end, the proof that strongly minimal sets in differentially closed fields are Zariski structures is given.

- *Differentially closed fields of characteristic zero:* a survey of basic results as well as more recent and sophisticated ones in the form they are needed in the other chapters. This contains mainly only indications of proofs as good references exist and are given.

- *Separably closed fields:* this chapter puts in place the setting for the characteristic p case of the conjecture. The first part consists of a short survey of the basic classical results on separably closed fields, for which again good references exist. The second part presents in detail, with all proofs given, the adequate topology for which dimension one subsets are Zariski geometries. The presentation and the proof given here by F. Delon are different from the ones given by Hrushovski in his paper. In particular the proof given here applies to all minimal types, removing the hypothesis of "thin" appearing in Hrushovski's account.

- *Proof of the Mordell-Lang conjecture:* a complete exposition of Hrushovski's proof in characteristic zero is given, the setting for the characteristic p case is set up in details and the main technical difficulties are pointed out. This chapter refers only to notions and results introduced in the previous chapters, except when mentioning the aspects of the characteristic p case which is not given in full detail here.

- *Proof of Manin's theorem by reduction to positive characteristic:* a proof is given of how to deduce the characteristic zero case from the positive characteristic case.

Prerequisites and notation

It will be assumed in all except the first chapter that the reader is familar with the very basic notions of classical model theory : models, formulas, definable sets, compactness theorem, saturation, types. All these notions are defined informally in the first chapter.

From the algebraic point of view, only a very basic knowledge of commutative algebra is assumed. All notions of algebraic geometry are defined, even the most basic ones.

Since these notes were written by different authors, they may in some instances use different terminology or notation, more adapted to their specific subject matter, but we have tried to signal these variations.

Otherwise we all use the standard notation of model theory, most of which is recalled in the first two chapters.

Acknowledgments

I would like first of all to thank all the contributors to this book. They showed great good will when asked again and again to make changes in order to converge towards a coherent whole. Some of them agreed to act as "help referees" for the others' chapters. Finally they all showed infinite patience over the years since this project was first initiated, years made so numerous only by my own procrastination.

Over these numerous years many others were called upon to contribute by comments and suggestions on parts of the book. I cannot thank them all but let me mention L. Bélair, B. Herwig, and give particular thanks to Z. Chatzidakis who without authoring directly any chapter, made valuable contributions to several, going well beyond the work any referee would do.

Finally, this project would not have been contemplated without the first step, the Manchester RESMOD workshop. This was masterfully organised at the Mathematics Department of the University of Manchester under the supervision of M. Prest.

Elisabeth Bouscaren

List of contributors

Elisabeth Bouscaren
Université Paris 7, UFR de Mathématiques, CNRS UPRESA 7056, 2 place Jussieu, 75251 Paris Cedex 05, France.

Françoise Delon
Université Paris 7, UFR de Mathématiques, CNRS UPRESA 7056, 2 place Jussieu, 75251 Paris Cedex 05, France.

Marc Hindry
Université Paris 7, UFR de Mathématiques, 2 place Jussieu, 75251 Paris Cedex 05, France.

Ehud Hrushovski
Department of Mathematics, Hebrew University, Givat Ram, Jerusalem, Israel.

Daniel Lascar
Université Paris 7, UFR de Mathématiques, CNRS UPRESA 7056, 2 place Jussieu, 75251 Paris Cedex 05, France.

David Marker
Department of Mathematics, University of Illinois at Chicago, 851 S.Morgan Street, Chicago, IL 60607-7045, USA.

Anand Pillay
Department of Mathematics, University of Illinois, Altgeld Hall, 1409 W. Green St., Urbana, IL 61801, USA.

Carol Wood
Department of Mathematics, Wesleyan University, Middletown, CT 06459, USA.

Martin Ziegler
Mathematisches Institut, Universität Freiburg, Eckerstr. 1, 79104 Freiburg, Germany.

Contents

Introduction to model theory

ELISABETH BOUSCAREN

In this informal presentation we introduce some of the main definitions and results which form the basis of model theory. We have chosen an approach adapted to the particular subject of this book. For proofs and formal definitions as well as for all that we have here purposely omitted, we suggest [Ho] or [Po 85] both rather close in spirit to the point of view adopted here. For a more classical approach, see [ChKe].

We illustrate all the notions we present with the three classes of fields which are at the center of our subject: algebraically closed fields, separably closed fields and differentially closed fields. In each case, a whole chapter of this book is devoted to the presentation of the model theory of these fields ([Pil], [De], [Wo]).

1 Structures, language associated to a structure

We begin with a first approach which has the advantage of focusing directly on the objects we are really interested in, the definable sets.

Definition 1.1 First definition of a structure
A **relational structure** $\mathcal{M} = \langle M, (B_i)_{i\in I}\rangle$ *consists of a (non empty) set* M, *and a family* $(B_i)_{i\in I}$ *of subsets of* $\bigcup_{n\geq 1} M^n$, *that is, for each* i, B_i *is a subset of* M^{n_i} *for some* $n_i \geq 1$. *We add the extra condition that the diagonal of* M^2 *is one of the* B_i*'s.*
Each B_i *is called an* **atomic subset** *of* $\mathcal{M}$.

Definition 1.2 *Let* $\mathcal{M} = \langle M, (B_i)_{i\in I}\rangle$ *be a relational structure. We define the family of* **definable subsets** *of the structure* $\mathcal{M}$, *denoted by* $Def(\mathcal{M})$.
$Def(\mathcal{M})$ *is the smallest family of subsets of* $\bigcup_{n\geq 1} M^n$ *with the following properties:*

1. *For every* $i \in I$, $B_i \in Def(\mathcal{M})$
2. $Def(\mathcal{M})$ *is closed under finite boolean combinations, i.e. if* $A, B \subseteq M^n$, $A, B \in Def(\mathcal{M})$, *then* $A\cup B \in Def(\mathcal{M})$, $A\cap B \in Def(\mathcal{M})$ *and* $M^n \setminus A \in Def(\mathcal{M})$.
3. $Def(\mathcal{M})$ *is closed under cartesian product, i.e. if* $A, B \in Def(\mathcal{M})$, $A \times B \in Def(\mathcal{M})$.

4. *Def*$(\mathcal{M})$ *is closed under projection, i.e. if* $A \subset M^{n+m}$, $A \in Def(\mathcal{M})$, *if* $\pi_n(A)$ *is the projection of* A *on* M^n, $\pi_n(A) \in Def(\mathcal{M})$.

5. *Def*$(\mathcal{M})$ *is closed under specialization, i.e. if* $A \in Def(\mathcal{M})$, $A \subseteq M^{n+k}$ *and if* $\overline{m} \in M^n$ *then*

$$A(\overline{m}) = \{\bar{b} \in M^k \ ; \ (\overline{m}, \bar{b}) \in A\} \ \in Def(\mathcal{M}).$$

6. *Def*$(\mathcal{M})$ *is closed under permutation of coordinates, i.e. if* $A \in Def(\mathcal{M})$, $A \subseteq M^n$, *if* σ *is any permutation of* $\{1, \ldots, n\}$,

$$\sigma(A) = \{(a_{\sigma(1)}, \ldots, a_{\sigma(n)}) \ ; \ (a_1, \ldots, a_n) \in A\} \ \in Def(\mathcal{M}).$$

Examples:

1. **Algebraically closed fields**
   Consider the following structure $\mathcal{K}$: K is an arbitrary algebraically closed field, and for the family $(B_i)_{i \in I}$ of atomic sets, take solutions to one polynomial equation, that is, sets of the form $\{\bar{a} \in K^n; P(\bar{a}) = 0\}$, for some $P(\overline{X}) \in K[X_1, \ldots, X_n]$. Closing under finite intersections, we get the Zariski closed subsets of K^n (or affine algebraic subsets of K^n), for all $n \geq 1$.

   Then if we close under boolean combinations, we get the *constructible* sets. It is well-known that constructible sets are closed under projection (this is Chevalley's theorem, or from the point of view of model theory, Tarski's elimination of quantifiers for algebraically closed fields). This means exactly that the definable sets of $\mathcal{K}$ are the constructible sets.

2. **Separably closed fields**
   Let S be a separably closed field of characteristic $p > 0$ and of finite (> 0) degree of imperfection. Take for the structure $\mathcal{S}$, the field S together with the same atomic sets as above, that is, zero sets of a polynomial in $S[X_1, \ldots, X_n]$, for every $n \geq 1$. Again, we can consider first the boolean combinations of atomic sets, that is the constructible sets. But because S is not algebraically closed, this family is no longer closed under projection, for example, the subfield $S^p = \{a \in S \ ; \text{ there is } b \in S, b^p = a \}$ is the projection of a constructible set, hence a definable set, but is not a constructible set.

3. **Differentially closed fields of characteristic** 0
   Let F be a differential field of characteristic zero, that is a field F with a derivation, a map $\delta : F \mapsto F$, such that $\delta(x + y) = \delta(x) + \delta(y)$ and $\delta(xy) = x\delta(y) + \delta(x)y$.

   In analogy with algebraically closed fields, one can define differentially closed fields: F is differentially closed if every finite set of differential equations and inequations which has a solution in some extension of F has a solution in F. In particular a differentially closed field is algebraically closed. Differentially closed fields exist and every differential field has a differential closure which is unique up to isomorphism.

For F a differential field, the **differential polynomial ring** over F in one differential indeterminate y is the ring $F\{y\} = F[y, \delta y, \delta^2 y, \ldots]$, i.e., the (usual algebraic) polynomial ring over F in the infinite set of indeterminates $y, \delta y, \delta^2 y, \ldots$ Similarly we denote the differential polynomial ring in differential indeterminates $y_1 \ldots, y_n$ by $F\{y_1, \ldots, y_n\}$.

We say that D is a ***principal δ-algebraic subset*** of F^n if
$D = \{(a_1, \ldots, a_n) \in F^n \,; P(a_1, \delta(a_1), \delta^2(a_1), ..., a_n, \delta(a_n), \delta^2(a_n), ...) = 0\}$
for some $P \in F\{y_1, \ldots, y_n\}$.

Now suppose that F is differentially closed and consider the structure $\mathcal{F}$ on F, where the atomic sets are the principal δ-algebraic sets. Call a set *δ-constructible* if it is a boolean combination of δ-algebraic sets. Then, as in the case of algebraicaly closed fields, one can prove that the projection of a δ-constructible set is a δ-constructible set (i.e. differentially closed fields have quantifier elimination). Hence the definable subsets are exactly the δ-constructible sets.

4. **Real closed fields**
Consider R any real closed field, or more specifically consider $\mathbb{R}$ the reals. The natural structure $\mathcal{R}$ associated to the reals is the following: take for the class of atomic sets the sets of the following form, for $P(\overline{X}) \in \mathbb{R}[\overline{X}]$:

$$\{\bar{a} \in \mathbb{R}^n \,; P(\bar{a}) = 0\} \text{ or } \{\bar{a} \in \mathbb{R}^n \,; P(\bar{a}) > 0\}.$$

Now closing by boolean combinations, one obtains the *semi-algebraic sets*. It is again well-known (Tarski-Seidenberg Theorem) that semi-algebraic sets are closed under projection. For a survey of the model theory of real closed fields, which we will only use as an example in this introduction, see [Mar 96].

This definition directly in terms of subsets is very natural but is not very useful. Indeed, the kind of natural questions we want to study are the following: consider the three field extensions

$$\mathbb{Q} \subset \widetilde{\mathbb{Q}} \subset \mathbb{C}$$

where $\widetilde{\mathbb{Q}}$ denotes the algebraic closure of $\mathbb{Q}$. Define the atomic subsets in each of these fields as above in cases 1 and 2, to be solutions to a polynomial. It is natural to study the relationship between $Def(\mathbb{Q})$, $Def(\widetilde{\mathbb{Q}})$ and $Def(\mathbb{C})$. For example, is there a canonical way to extend a definable subset of $\mathbb{Q}$ to a definable subset of $\widetilde{\mathbb{Q}}$, of $\mathbb{C}$? Conversely if X is a definable subset in $\mathbb{C}^n$, what about $X \cap \widetilde{\mathbb{Q}}^n$, $X \cap \mathbb{Q}^n$?

Any countable field of characteristic zero can be embedded into $\mathbb{C}$. Do such "universal" structures exist more generally for other classes of structures? The well known "Lefschetz Principle" tells us that any "algebraic" property true in a algebraically closed field of given characteristic is true in any other algebraically closed field of the same characteristic. Can we formalize such a principle for other classes of structures? Given two structures, how similar or different can they be, what is the natural notion of a homomorphism, of an isomorphism?

In order to deal with this type of questions, we need a more "constructive" way to describe the definable sets which does not depend on the chosen structure. We will then define the definable sets as the sets of solutions, in a given structure, of a "formula", in the same way one considers, in algebraic geometry, the set of rational points in a given field of an affine algebraic set or of a variety defined over this field.

Definition 1.3 • *A* **structure** $\mathcal{M}$ *consists of*

1. *A non empty set M.*
2. *A family $(R_i{}^M)_{i\in I}$ of relations or subsets of $\bigcup_{n\geq 1} M^n$, that is, for each i, $R_i{}^M$ is a subset of M^{n_i} for some $n_i \geq 1$. We add the extra condition that the diagonal of M^2 is one of the $R_i{}^M$'s.*
   For convenience we will allow ourselves the two notations $(a_1,\ldots,a_{n_i}) \in R_i{}^M$ or $R_i{}^M(a_1,\ldots,a_{n_i})$.
3. *A family of maps $(f_j{}^M)_{j\in J}$, where $f_j{}^M$ is an n_j-ary map, $f_j{}^M : M^{n_j} \mapsto M$.*
4. *A set of constants $(c_k{}^M)_{k\in K}$, where $c_k{}^M \in M$.*

• *The* **signature** *or* **language** *L associated to a structure $\mathcal{M}$ consists of*

1. *For each relation $R_i{}^M$, a relation symbol, R_i of arity n_i. As the diagonal of M^2 is by assumption always one of our basic relations, there always is a particular binary relation in L, which corresponds to equality in the structure, and which we allow ourselves not to mention when describing a specific language.*
2. *For each map $f_j{}^M$ a function symbol , f_j of arity n_j.*
3. *For each constant $c_k{}^M$ a constant symbol c_k.*

We will use the following notation for an arbitrary language

$$L = \{(R_i)_{i\in I}; (f_j)_{j\in J}; (c_k)_{k\in K}\}.$$

• *If L is the language (or signature) associated to the structure $\mathcal{M}$, we say that $\mathcal{M}$ is an* **L-structure**.

Note that a language need not be countable.

Examples: Let us go back to our previous examples

1. In the case of algebraically closed fields and of separably closed fields, we will consider the signature or language natural for rings, L_{ring} consisting of 2 constants, 0 and 1, two binary maps, one for addition and one for multiplication, one unary map for the inverse in the additive group; $L_{ring} = \{+; .; -; 0; 1\}$.
2. In the case of differential fields, we add to L_{ring} a new unary map, which will be interpreted by the derivation map δ; $L_{diff} = \{+; .; -; \delta, 0; 1\}$.
3. In the case of real closed fields, or more generally of any ordered ring R, we add to the language L_{ring} a new binary relation, interpreted by the order relation on R; $L_{ord} = \{<; +; .; -; 0; 1\}$.

We are here allowing ourselves to use the same notation for the symbol in the language and the actual relation, map or element of the structure that it represents.

Now before we explain how to construct the atomic sets and the definable sets from the language, we define the notion of homomorphism.
Let $L = \{(R_i)_{i \in I}; (f_j)_{j \in J}; (c_k)_{k \in K}\}$ be a language, let $\mathcal{M}$ and $\mathcal{N}$ be two L-structures.

1. A map h from M to N is a **homomorphism** from $\mathcal{M}$ to $\mathcal{N}$ if the following holds
   - for every constant $c \in L$, $h(c^M) = c^N$
   - for every n-ary function $f \in L$, for every $\bar{a} \in M^n$, $h(f^M(\bar{a})) = f^N(h(\bar{a}))$
   - for every n-ary relation $R \in L$, for every $\bar{a} \in M^n$, if $\bar{a} \in R^M$, then $h(\bar{a}) \in R^N$.
2. A homomorphism h from $\mathcal{M}$ to $\mathcal{N}$ is an **embedding** if, for all n-ary relation $R \in L$, for every $\bar{a} \in M^n$, $\bar{a} \in R^M$ **iff** $\bar{a} \in R^N$.
   Note that, as the diagonal is always in L, an embedding is injective.
3. An **isomorphism** from $\mathcal{M}$ to $\mathcal{N}$ is a surjective embedding. An automorphism is an isomorphism from $\mathcal{M}$ onto itself.
4. $\mathcal{M}$ is an **L-substructure** of $\mathcal{N}$, denoted $\mathcal{M} \subseteq_L \mathcal{N}$, if $M \subseteq N$ and the inclusion map from M into N is an L-embedding.
   This is equivalent to the following conditions:
   - for every constant $c \in L$, $c^M = c^N$
   - for every n-ary function $f \in L$, for every $\bar{a} \in M^n$, $f^M(\bar{a}) = f^N(\bar{a}) \in M$
   - for every n-ary relation $R \in L$, $R^M = R^N \cap M^n$.

The notion of substructure depends on the choice we have made of a language to describe a given structure. For example, given a field K, an L_{ring}-substructure of A will be a subring of K. We could have chosen to add an extra unary map to the language, to be interpreted by the inverse of multiplication on K^* (and taking value 0 on 0 for example). In that case a substructure would be a subfield of K. We usually choose a language appropriate or natural for the questions we want to consider.

2 Definable sets and formulas, satisfaction

Warning: We will now explain how to retrieve the definable sets from the language. We will give the basic steps of the construction but the reader should be aware that there are some very basic technical (but unavoidable) points that we will ignore here, which need to be dealt with in order for this construction to be correct and coherent and for the expressions or objects we introduce to be uniquely defined.

We have a fixed language L, we give ourselves an infinite set of variables which we will usually denote by $v, w, x, y, x_i, \ldots$.

We first define the set of **terms** of the language L. Terms allow us to compose the maps from our language, and also to define new elements from the given constants. They are generated inductively by the following rules:

1. every variable is a term
2. every constant of L is a term
3. if $f \in L$ is an n-ary function, if $t_1, \ldots, t_n$ are terms, then $f(t_1, \ldots, t_n)$ is a term

Example: In the case of the language of rings, $L_{ring} = \{0, 1, +, ., -\}$, or of the language for ordered rings $L_{ord} = \{0, 1, +, ., -, <\}$, a term is exactly a polynomial $P(x_1, \ldots, x_n)$ in $\mathbb{Z}[x_1, \ldots, x_n]$ for some n.
In the case of the language for differential rings $L_{diff} = \{0, 1, +, ., -, \delta\}$, a term is exactly a differential polynomial in the ring $\mathbb{Z}\{X_1, \ldots, X_n\}$ for some n.

Now we define the set of **atomic formulas** of L which give rise to what we called the atomic subsets.
An atomic formula of L is an expression of the form: $R(t_1, \ldots, t_n)$ where R is an n-ary relation in L and $t_1, \ldots, t_n$ are terms of L.
Note that as equality is always in L, for any two terms s and t, "$s = t$" is an atomic formula.
Examples: In the case of L_{ring} the atomic formulas are formulas of the form "$P(x_1, \ldots, x_n) = Q(x_1, \ldots, x_n)$", where $P, Q \in \mathbb{Z}[x_1, \ldots, x_n]$.
In the case of L_{diff}, we get equations of differential polynomials, and in the case of L_{ord} we get additional formulas of the form "$P(x_1, \ldots, x_n) < Q(x_1, \ldots, x_n)$".

Now given $\mathcal{M}$ an L-structure, we want to define what it means for an atomic formula to be satisfied in $\mathcal{M}$.

Satisfaction of atomic formulas and atomic subsets

Let us deal first with the case when we have an atomic formula ϕ, with no variables, for example in L_{ring}, "$0 = 1$" or more interestingly,

$$\text{"}\underbrace{1 + \ldots + 1}_{p} = 0\text{"} \ (p.1 = 0).$$

Then in a given L_{ring}-structure such a formula will be either true or false, in particular "$p.1 = 0$" will be true in a ring iff it has characteristic p.
Now if $\phi(x_1, \ldots, x_n)$ is an atomic formula with n variables, then it is of the form $R(t_1(x_1, \ldots, x_n), \ldots, t_k(x_1, \ldots, x_n))$ for R some k-ary relation in L, and the $t_i(x_1, \ldots, x_n)$'s terms of L with variables amongst $x_1, \ldots, x_n$. Given an n-tuple $\bar{a} = (a_1, \ldots, a_n) \in M^n$, we say that $(a_1, \ldots, a_n)$ **satisfies** $\phi(x_1, \ldots, x_n)$ in $\mathcal{M}$ if $R^M(t_1(\bar{a}), \ldots, t_k(\bar{a}))$. We denote this by

$$\mathcal{M} \models \phi(a_1, \ldots, a_n).$$

If the tuple $(a_1, \ldots, a_n)$ does not satisfy $\phi(x_1, \ldots, x_n)$ in $\mathcal{M}$, we write

$$\mathcal{M} \not\models \phi(a_1, \ldots, a_n).$$

Now we say that $S \subseteq M^n$ is an **atomic subset** if

$$S = \{(a_1, \ldots, a_n) \in M^n \ ; \ \mathcal{M} \models \phi(a_1, \ldots, a_n, b_1, \ldots, b_m)\}$$

for some atomic formula $\phi(x_1, \ldots, x_n, x_{n+1}, \ldots, x_{n+m})$ and some $\bar{b} \in M^m$. We say that S is defined with parameters $\bar{b}$ or defined over $\bar{b}$.

In this way we retrieve in our previous examples, our original atomic subsets.
Note that again we are ignoring problems having to do with permutation of variables, the way in which we list the variables in a formula etc. These have to be dealt with of course.

Formulas and definable sets

We now want to generate the definable sets from the atomic sets, or seen from the point of view of the language, the formulas from the atomic formulas. This will be done inductively, generating new formulas with the following operators: $\wedge$ (and) which corresponds to intersection, $\vee$ (or) which corresponds to union, $\neg$ (not) which corresponds to complementation, $\exists$ (there exists) corresponding to projection and $\forall$ (for all).

We are going to define simultaneously the **set of formulas** and **satisfaction of a formula ϕ in an L-structure** $\mathcal{M}$, denoted as before $\mathcal{M} \models \phi$. All of this is the formal translation of the obvious clauses, we write them out partially in order to fix notation.

- All atomic formulas are formulas.
- If $\phi_1(x_1, \ldots, x_n)$ and $\phi_2(x_1, \ldots, x_n)$ are formulas, then $(\phi_1 \wedge \phi_2)(x_1, \ldots, x_n))$ is a formula.
  If $\bar{a} \in M^n$, $\mathcal{M} \models (\phi_1 \wedge \phi_2)(\bar{a})$ iff $\mathcal{M} \models \phi_1(\bar{a})$ and $\mathcal{M} \models \phi_2(\bar{a})$.

- If $\phi_1(x_1, \ldots, x_n)$ is a formula, $(\neg\phi_1(x_1, \ldots, x_n))$ is a formula.
  If $\bar{a} \in M^n$, $\mathcal{M} \models \neg\phi_1(a_1, \ldots, a_n)$ iff $\mathcal{M} \not\models \phi_1(a_1, \ldots, a_n)$.

- If $\phi_1(x_1, \ldots, x_n)$ is a formula, then $\exists x_n\ \phi_1(x_1, \ldots, x_n)$ is a formula.
  If $(a_1, \ldots, a_{n-1}) \in M^{n-1}$, $\mathcal{M} \models \exists x_n\ \phi_1(a_1, \ldots, a_{n-1}, x_n)$ iff there exists $b \in M$ such that $\mathcal{M} \models \phi_1(a_1, \ldots, a_{n-1}, b)$.

- The obvious clauses also for $\vee$ and $\forall$.

By abuse of notation we also denote by L the set of formulas of the language L. Note that the cardinality of the set of formulas for the language L, denoted $|L|$, is equal to the supremum of the cardinality of the language and $\aleph_0$.

Now we say that $D \subseteq M^n$ is a **definable subset** in the L-structure $\mathcal{M}$ if there are $\bar{b} \in M^m$ ($\bar{b}$ may be empty) and a formula $\phi(x_1, ..., x_n, x_{n+1}, ..., x_{n+m})$ such that

$$D = \{(a_1, \ldots, a_n) \in M^n \ ; \ \mathcal{M} \models \phi(a_1, ..., a_n, b_1, ..., b_m)\}.$$

We say that D is definable with parameters B (or over B) or defined by a formula with parameters in B if $\bar{b} \subseteq B$.
Note that if $A \subseteq M$, then the formulas with parameters in A, $\{\phi(\bar{x}, \bar{a}); \bar{a} \in A, \phi(\bar{x}, \bar{y}) \in L\}$ are exactly the formulas of the language $L(A) = L \cup \{c_a; a \in A\}$, where one adds to the original L a new constant for each element of A.

It should be clear that the definable sets in this sense are exactly $Def(M)$ for the relational structure $\langle M, (A_i)_{i \in I}\rangle$ where the family (A_i) is the family of all the atomic definable sets.

A family $(E_i)_{i \in I}$ of definable subsets of M^n is **uniformly definable** if there is a formula $\phi(\bar{x}, \bar{y}) \in L$ and $(\bar{b}_i)_{i \in I}$, $\bar{b}_i \in M^k$ such that for each i, $E_i = \{\bar{a} \in M^n; M \models \phi(\bar{a}, \bar{b}_i)\}$. In an algebraically closed field K, an algebraic family of affine subsets of K^n is an example of a uniformly definable family. Another typical example is that of the family of centralizers of elements in a group: consider a group G as an L_{group}-structure, where L_{group} is one of the possible natural languages for groups, e.g. $L_{group} = \{., e, ^{-1}\}$. Then the family $(C(b))_{b \in G}$, where $C(b)$ denotes the centralizer of b in G, is uniformly definable, with the formula $\phi(x, y)$: "$x.y = y.x$".

3 Theories, elementary substructures, compactness

We now want to formalize the idea of structures being "similar". This is what elementary equivalence and theories enable us to do.

We say that a variable x is free in a formula ϕ if it is not in the scope of a quantifier ($\exists$ or $\forall$). When we write a formula as $\phi(x_1, \ldots, x_n)$ we mean that the free variables in the formula ϕ are amongst $x_1, \ldots, x_n$. A formula with no free variables is called a **sentence**. Note that, given an L-structure M, a sentence σ of L is either satisfied (true) in M or not satisfied (false) in M. In this second case, the negation of σ, the sentence $\neg\sigma$ is then true in M.

A set Σ of sentences in L is **consistent** (or satisfiable) if there is an L-structure $\mathcal{M}$ such that for every sentence $\sigma \in \Sigma$, $\mathcal{M} \models \sigma$. In this case, we say that $\mathcal{M}$ is a **model** of Σ, denoted $\mathcal{M} \models \Sigma$.

The usual axioms for commutative rings are obviously expressible as sentences in the language L_{ring} (for example, we say that addition is a commutative group operation by: $\forall x\, \forall y\; x + y = y + x$, $\forall x\; x + 0 = x$, $\forall x\, \forall y\, \forall z\; x + (y + z) = (x + y) + z$, $\forall x\; x + (-x) = 0$, and so on for multiplication and distributivity). Let us call this finite set of sentences Σ_r. An L_{ring}-structure A is a commutative ring iff A satisfies Σ_r; a commutative ring A is a field iff it also satisfies σ: $\forall x\; (x = 0) \lor (\exists y\; x.y = 1)$.

If we add the infinite set of sentences

$$\Sigma_{alg} = \{\forall y_0 \ldots \forall y_{n-1}\, \exists x\; x^n + \sum_{i=0}^{n-1} y_i.x^i = 0\; ;\; n \geq 1\},$$

then $A \models \Sigma_r \cup \{\sigma\} \cup \Sigma_{alg}$ if and only if A is an algebraically closed field.

A **theory** in L is a consistent set of sentences in L. A theory T in L is said to be **complete** if it is maximal, that is if for any sentence σ in L, $\sigma \in T$ or $\neg\sigma \in T$ (of course at most one of σ and $\neg\sigma$ can be a member of a consistent set of sentences).

Theories are a first step towards classifying structures: given a class K of L-structures, we can consider the associated theory, consisting of all sentences which are true in all structures in K. If we consider a single L-structure $\mathcal{M}$, then the associated theory, $Th(\mathcal{M}) = \{\sigma \in L; \mathcal{M} \models \sigma\}$, is always complete.

More interestingly, if we consider the theory of all algebraically closed fields of characteristic p (for $p \geq 0$), that is the set of sentences in L_{ring} which are true in all algebraically closed fields of characteristic p, then this is a complete theory. Similarly, the theory in L_{diff} of all differentially closed fields of characteristic zero and the theory in L_{ord} of all real closed fields are both complete.

This means that for example any two algebraically closed fields of same characteristic satisfy exactly the same sentences. More generally, given two L-structures $\mathcal{M}$ and $\mathcal{N}$ we say that $\mathcal{M}$ and $\mathcal{N}$ are **elementarily equivalent**, denoted $\mathcal{M} \equiv \mathcal{N}$, if for all sentence $\sigma \in L$, $\mathcal{M} \models \sigma$ iff $\mathcal{N} \models \sigma$.

Given $\mathcal{M} \subseteq_L \mathcal{N}$, it is not enough to suppose that $\mathcal{M} \equiv \mathcal{N}$ in order to deal with some of the questions we are interested in concerning the definable subsets of $\mathcal{M}$ and $\mathcal{N}$.

Definition 1) Let $\mathcal{M} \subseteq_L \mathcal{N}$. We say that $\mathcal{M}$ is an **elementary substructure** of $\mathcal{N}$ (or that $\mathcal{N}$ is an elementary extension of $\mathcal{M}$), denoted by $\mathcal{M} \preceq \mathcal{N}$, if for all formula $\phi(x_1, \ldots, x_n) \in L$, for all $\bar{a} = a_1, \ldots, a_n \in M^n$,

$$\mathcal{M} \models \phi(a_1, \ldots, a_n) \text{ iff } \mathcal{N} \models \phi(a_1, \ldots, a_n).$$

Note that this holds in particular for sentences of L, hence $\mathcal{M}$ and $\mathcal{N}$ are elementarily equivalent.
2) A map f from M to N is an **elementary embedding** from $\mathcal{M}$ into $\mathcal{N}$ if for all formula $\phi(x_1, \ldots, x_n) \in L$, for all $\bar{a} \in M^n$,

$$\mathcal{M} \models \phi(a_1, \ldots, a_n) \text{ iff } \mathcal{N} \models \phi(f(a_1), \ldots, f(a_n)).$$

This means exactly that $\mathcal{M}$ is isomorphic to an elementary substructure of $\mathcal{N}$.

All embeddings are of course not elementary, but the following is important to note and is proved easily by induction on formulas:

Fact 3.1 *Any isomorphism is an elementary embedding.*

If $M \preceq N$, then any definable subset S of M^n has a canonical extension to a definable subset S' of N^n, such that $S' \cap M^n = S$:

$$\text{if } S = \{\overline{m} \in M; \mathcal{M} \models \phi(\overline{m})\} \text{ then } S' = \{\bar{n} \in N^n; \mathcal{N} \models \phi(\bar{n})\}.$$

In practice the following criterium is the most useful to describe elementary extensions in terms of definable sets.

Fact 3.2 (Tarski-Vaught test) *Let $\mathcal{M} \subseteq_L \mathcal{N}$. Then $\mathcal{M} \preceq \mathcal{N}$ iff for all formula $\phi(x, y_1, \ldots, y_n) \in L$, for all $m_1, \ldots, m_n \in M$, such that $\mathcal{N} \models \exists x\, \phi(x, m_1, \ldots, m_n)$, there is some $m_0 \in M$ such that $\mathcal{N} \models \phi(m_0, m_1, \ldots, m_n)$.*

This means that $\mathcal{M} \preceq \mathcal{N}$ iff for all (non empty) definable subset $E \subseteq N$, defined with parameters from M, $E \cap M$ is non empty. It follows that the same is true for definable subsets of N^n.

Let us go back now to some of our examples.

- Fix $p \geq 0$. Let ACFp denote the (complete) theory of all algebraically closed fields of characterisitic p (in the language L_{ring}). We have seen that being algebraically closed is expressible by an infinite set of sentences, and being of characterisitic p is also expressible. Hence the models of ACFp are exactly all the algebraically closed fields of characteristic p. The fact that the projection of a constructible set is again constructible means that the theory ACFp has **quantifier elimination** : in any model of ACFp, any formula is equivalent to a formula without quantifiers, i.e. to a boolean combination of atomic formulas. It follows that any embedding between two algebraically closed fields of characteristic p is elementary.
- As we remarked at the beginning, the theory of separably closed fields does not have quantifier elimination in the language L_{ring}. We will see later in this volume that by suitably enlarging the language one easily achieves quantifier elimination.
- In the language L_{ord}, real closed fields have elimination of quantifiers (because the projection of a semi-algebraic set is again a semi-algebraic set). We could

have chosen to consider the theory of real closed fields in the language L_{ring}, without the ordering, as the ordering is definable in the language L_{ring}:

$$x < y \text{ iff } \exists z\ (z^2 + x = y \wedge \neg(z = 0)).$$

These two theories, the one in L_{ring} and the one in L_{ord} have exactly the same models, but it is not possible to eliminate quantifiers in the smaller language L_{ring}. In particular the formula given above to define the order relation is not equivalent to a formula without quantifiers in L_{ring}.
- Consider again the field extensions (in L_{ring})

$$\mathbb{Q} \subset \widetilde{\mathbb{Q}} \subset \mathbb{C}$$

We have that $\widetilde{\mathbb{Q}} \preceq \mathbb{C}$. Certainly $\mathbb{Q} \subset_{L_{ring}} \widetilde{\mathbb{Q}}$; it follows that if $\phi(\bar{x})$ is a formula with no quantifier, if $\bar{a} \in \mathbb{Q}^n$, then

$$\mathbb{Q} \models \phi(\bar{a}) \text{ iff } \widetilde{\mathbb{Q}} \models \phi(\bar{a})$$

But we do not have that $\mathbb{Q} \preceq \widetilde{\mathbb{Q}}$, they are not even elementarily equivalent : for example $\mathbb{Q} \not\models \exists x\ x^2 = -1$.

One of the fundamental theorems of basic model theory is the Compactness theorem, which corresponds to the fact that complete theories form the Stone Space of an adequate Boolean Algebra (see section 4). For the sake of clarity we separate the statement of the Compactness theorem in two cases.

Fact 3.3 (Compactness theorem). **1.** *Let Σ be a set of sentences in L. Then Σ is consistent (i.e. has a model) iff every finite subset of Σ is consistent (has a model).*
2. *Let $\mathcal{M}$ be an L-structure and let $\Sigma(x_1, \ldots, x_n) = \{\phi_i(x_1, \ldots, x_n); i \in I\}$ be a set of formulas in L with n variables (allowing parameters from M). Suppose that for each finite subset $\{i_1, \ldots, i_k\}$ of I,*

$$\mathcal{M} \models \exists x_1 \ldots \exists x_n\ (\phi_{i_1}(x_1, \ldots, x_n) \wedge \ldots \wedge \phi_{i_k}(x_1, \ldots, x_n)).$$

Then there is some $\mathcal{N} \succeq M$ and some $\bar{a} \in N^n$ such that for every $i \in I$ $\mathcal{N} \models \phi_i(\bar{a})$.

In terms of definable sets, this means that if we have an infinite family $\mathcal{F} = \{D_i; i \in I\}$ of definable subsets of M^n which has the finite intersection property in M (for any finite $J \subset I$, $\bigcap_{i \in J} D_i \neq \emptyset$), then in some elementary extension $\mathcal{N}$ of $\mathcal{M}$, the family $\mathcal{F}$ has non empty intersection ($\bigcap_{i \in I} D_i \neq \emptyset$). This may not be true in M itself though: consider the family $\mathcal{F}$ of all subsets of $\mathbb{Q}^n$ defined by $P(x_1, \ldots, x_n) \neq 0$ for some non zero polynomial $P(X_1, \ldots, X_n) \in \mathbb{Q}[X_1, \ldots, X_n]$. Any $a_1, \ldots, a_n$ in the intersection of $\mathcal{F}$ must be algebraically independent over $\mathbb{Q}$. The Compactness theorem tells us that for every n, $\mathbb{Q}$ has

an elementary extension of transcendence degree at least n.
We can also consider the reals in the language L_{ord}; in this case the Compactness theorem implies for example that $\mathbb{R}$ has a non archimedian elementary extension (i.e. an elementary extension with infinitely small and infinitely large elements).

The following theorem is also classical; statement 2. is a direct consequence of compactness. It says in particular that if L is countable and T is a theory with infinite models, then T has models in all cardinalities.
For A any set, $|A|$ denotes the cardinality of A.

Fact 3.4 (Löwenheim-Skolem theorem)
1. *Let L be a language, $\mathcal{M}$ an L-structure and $X \subseteq M$. Then there exists $M_0 \preceq M$ such that $X \subseteq M_0$ and $|M_0| \leq |X| + |L|$.*
2. *Let L be a language, let $\mathcal{M}$ be an infinite L-structure. Then for any cardinal $\kappa > |M|$, $\mathcal{M}$ has an elementary extension of cardinality κ.*

Now some models are richer than others in the sense that families of definable sets with the finite intersection property will have non empty intersection in them. This is for example the case in $\mathbb{C}$ for any countable family of definable sets. Such models always exist:

Definition 3.5 *Let κ be an infinite cardinal. The L-structure $\mathcal{M}$ is* **κ-saturated** *if for any family $\mathcal{F} = \{D_i; i < \kappa\}$ of definable subsets of M with the finite intersection property, there is some $a \in M$, $a \in \bigcap_{i<\kappa} D_i$.*
We say that $\mathcal{M}$ is **saturated** *if $\mathcal{M}$ is $|M|$-saturated.*

Note that in the above definition, we considered only definable subsets of M, that is, formulas in one variable. But in fact it follows that the same will be true for any family of less than κ definable subsets of M^n for every $n \geq 1$.
In the case of algebraically closed fields, it is easy to see that a field K is κ-saturated iff K has transcendence degree at least κ over the prime field.

Fact 3.6 **1.** *Let κ be an infinite cardinal, let $\mathcal{M}$ be an L-structure. Then there is a κ-saturated L-structure $\mathcal{N}$ such that $\mathcal{M} \preceq \mathcal{N}$.*
2. *Let κ be an infinite cardinal and let $\mathcal{M}$ be a κ-saturated L-structure. If $\mathcal{N}$ is an L-structure of cardinality less than κ and $\mathcal{N} \equiv \mathcal{M}$, there is an elementary embedding from $\mathcal{N}$ into $\mathcal{M}$.*

The first statement follows again from the compactness theorem; the second one says that κ-saturated structures correspond to the "universal domains" traditionally considered in algebra.
The κ-saturated models of cardinality κ (the saturated models) are particularly useful: they have the additional property of homogeneity.
We say that an L-structure $\mathcal{M}$ of cardinality κ is **homogeneous** if for any $A, B \subset M$, of cardinality less than κ, and any partial elementary embedding f from A onto B, f extends to an automorphism of M.

It is not true in general that any theory has saturated models of cardinality κ for every κ. Some of the obstructions to this will appear in the following section.
But it is possible to assume for convenience (one formal and correct way to do this is explained at the beginning of the next chapter of this book), that any complete theory has saturated models of arbitrarily large cardinality, and even more that we have at our disposition a "monster" model which is saturated and of cardinality much bigger than any of the structures we are dealing with.
This is a particularly innocuous assumption to make in the case of the classes of structures we will be interested in the rest of this book. Indeed we will consider only "stable" theories for which it is true without any additional condition that arbitrarily large saturated models exist.

There is one last indispensable notion we need to introduce here, the notion of type. This is the object of the last section.

4 Types

Let T be a complete theory in a language L. Fix a model $\mathcal{C}$ of T, saturated of cardinality κ for some very large cardinal κ (in particular κ is much larger than the cardinality of the language). **By convention** from now on, all models $\mathcal{M}, \mathcal{N} \ldots$ of T will be elementary substructures of the big model $\mathcal{C}$ of cardinality less than κ. All subsets $A, B, D \ldots$ we consider will be subsets of $\mathcal{C}$ of cardinality less than κ.

Let $A \subset \mathcal{C}$ and let $\Sigma(x_1, \ldots, x_n) = \{\phi_i(x_1, \ldots, x_n); i \in I\}$ be a set of L-formulas with n variables and parameters from A (or equivalently a set of $L(A)$-formulas with n variables). We say that $\Sigma(x_1, \ldots, x_n)$ is **finitely satisfiable** if for any finite subset F of $\Sigma(x_1, \ldots, x_n)$, there is some $(c_1, \ldots, c_n) \in \mathcal{C}^n$ such that

$$\mathcal{C} \models \phi(c_1, \ldots, c_n) \text{ for every } \phi(x_1, \ldots, x_n) \in F.$$

This is the same as taking a family of A-definable subsets of $\mathcal{C}^n$ with the finite intersection property.

A **complete n-type** over A is a finitely satisfiable set $\Sigma(x_1, \ldots, x_n)$ of formulas in n variables with parameters in A which is maximal, i.e. such that for every formula $\phi(x_1, \ldots, x_n)$ in $L(A)$, $\phi(x_1, \ldots, x_n) \in \Sigma(x_1, \ldots, x_n)$ or $\neg\phi(x_1, \ldots, x_n) \in \Sigma(x_1, \ldots, x_n)$.

By saturation of $\mathcal{C}$ (and our convention that A has cardinality less than κ), there must be some $(c_1, \ldots, c_n) \in \mathcal{C}^n$ such that $\mathcal{C} \models \phi(c_1, \ldots, c_n)$ for every $\phi(x_1, \ldots, x_n) \in \Sigma(x_1, \ldots, x_n)$.
We say that $(c_1, \ldots, c_n)$ **realizes** the complete type Σ over A and write $\mathcal{C} \models \Sigma(c_1, \ldots, c_n)$.

Conversely, if we consider any $(c_1, \ldots, c_n) \in \mathcal{C}^n$, then let the **type of** $(c_1, \ldots, c_n)$ **over** A, denoted by $t(c_1, \ldots, c_n/A)$, be the following set of formulas:

$$t(c_1, \ldots, c_n/A) = \{\phi(x_1, \ldots, x_n) \in L(A); \mathcal{C} \models \phi(c_1, \ldots, c_n)\}.$$

Then $t(c_1, \ldots, c_n/A)$ is a complete n-type over A. This corresponds to considering the family of all A-definable subsets of C^n which contain $(c_1, \ldots, c_n)$.

Before we go back to our examples, one more useful fact. If f is an automorphism of $\mathcal{C}$ fixing the set A pointwise, $f \in Aut_A(C)$, then certainly f must fix setwise every A-definable subset and hence must fix setwise all complete types over A. By homogeneity of the big model $\mathcal{C}$, the converse is true: let $\bar{c}, \bar{d} \in C^n$,

$$t(\bar{c}/A) = t(\bar{d}/A) \text{ iff there is some } f \in Aut_A(C) \text{ such that } f(\bar{c}) = \bar{d}.$$

In a saturated model, complete n-types over A correspond exactly to orbits of n-tuples under the action of automorphisms fixing A pointwise.

Examples:

1. **Algebraically closed fields.** We will see in the relevant chapter that if K is an algebraically closed field, if k is a subfield of K, there is a one-to-one correspondance between complete n-types over k and prime ideals in the polynomial ring $k[X_1, \ldots, X_n]$.
2. **Differentially closed fields of characteristic zero.** Similarly we will see that quantifier elimination for the theory of differentially closed fields of characteristic 0 in the language L_{diff} implies that if K is a differentially closed field and k is a differential subfield of K, there is a one-to-one correspondance between complete n-types over k and prime differential ideals in the differential polynomial ring $k\{X_1, \ldots, X_n\}$.
3. **Separably closed fields of characteristic p.** We will see that, by enlarging the language L_{ring} in a natural way, we also get a one-to-one correspondance between types and prime separable ideals in a suitable ring of polynomials.

More generally we will be naturally led to consider arbitrary intersections of definable sets.

Definition 4.1 *Let $A \subset C$. A* **partial type** *over A in C^n, also called an* **infinitely definable** *subset over A (also $\wedge\wedge$-definable or $\wedge$-definable or ∞-definable), is an infinite intersection of A-definable subsets of C^n.*

For example, in a separably closed field K of characteristic $p > 0$, the field $K^{p^\infty} = \bigcap_n K^{p^n}$ is not definable but it is infinitely definable.

In algebraically closed fields, there is a natural way to associate to a Zariski closed set, and more generally to an abstract variety, a dimension. One of the main topics of stability theory is to define generalized notions of dimension, which we call ranks, which can be associated to definable sets and to types. We will see in the next chapter ("Introduction to stability theory and Morley rank") a selfcontained and detailed exposition of how to do this for a certain class of theories which we call ω-stable (or totally transcendental). Algebraically closed fields and differentially closed fields are two examples of ω-stable theories.

We will see that the structure of the sets of complete types is very relevant in this context. Let us end this informal introduction by indicating the first basic information that can be derived from these spaces of types.

The space of types

Let L be a language, which from now on we assume to be **countable.** Let T be a complete theory in L and let as above $\mathcal{C}$ be a saturated model of T of very large cardinality. We continue with our convention that all sets and models of T that we consider are of cardinality strictly smaller than the cardinality of $\mathcal{C}$.

Let $A \subset \mathcal{C}$. Fix some $n \geq 1$ and consider $Def_A{}^n$ the collection of all A-definable subsets of $\mathcal{C}^n$. We will identify a given definable subset of $\mathcal{C}^n$ and the formula used to define it, $\phi(\bar{x}, \bar{a})$, where $\bar{a}$ is a tuple of elements from A (two different formulas may define the same subset but we work up to equivalence of formulas in $\mathcal{C}$ in the obvious sense) and $\bar{x} = (x_1, \ldots, x_n)$.

$Def_A{}^n$ is a boolean algebra for the usual operations of intersection, union and complementation. A complete n-type over A is exactly an ultrafilter in this boolean algebra. The **space of complete n-types over** A, denoted by $S_n(A)$ is the Stone space associated to the boolean algebra $Def_A{}^n$. We consider $S_n(A)$ with the (usual) topology where a basis of open sets is given by

$$\langle \phi(\bar{x}, \bar{a}) \rangle = \{p \in S_n(A); \phi(\bar{x}, \bar{a}) \in p\}$$

for all formulas $\phi(\bar{x}, \bar{y}) \in L$ and all finite $\bar{a} \subseteq A$.

Every such basic open set is also closed (the complement corresponds to another definable set) and the Compactness theorem (applied to subsets of the big saturated model $\mathcal{C}$) stated in the previous section corresponds to the fact that $S_n(A)$ is compact. Note that

$$\langle \phi(\bar{x}, \bar{a}) \rangle \neq \emptyset \text{ iff } \mathcal{C} \models \exists \bar{x}\ \phi(\bar{x}, \bar{a}).$$

We can already derive some consequences from the fact that $S_n(A)$ is compact totally disconnected and separable when A is countable.

In particular compact spaces have the Baire property: a subset E is said to be meager if it is contained in a countable union of closed sets with empty interior, compact spaces are not meager. From the Baire property one can deduce the important Omitting types theorem:

Definition 4.2 *Let $p \in S_n(A)$ and let $\mathcal{M}$ be a model of T, $A \subset M$. We say that $\mathcal{M}$* **omits** *the type p if p is not realized in M^n.*

Note that a type $p \in S_n(A)$ is isolated iff there is a definable set $\phi(\bar{x}, \bar{a})$ such that, for any $\bar{c} \in \mathcal{C}^n$, $t(\bar{c}/A) = p$ iff $\mathcal{C} \models \phi(\bar{c}, \bar{a})$. It follows that in any model $\mathcal{M} \preceq \mathcal{C}$ containing A, the type p must be realized. But for example the field $\widetilde{\mathbb{Q}}$ omits the type of transcendental elements, which is not isolated.

Theorem 4.3 (Omitting types theorem) *Let L be a countable language and T a complete theory in L. For each n, let X_n be a meager subset of $S_n(\emptyset)$. Then there is a countable model $\mathcal{M}$ of T which omits every type in $\bigcup_{n\geq 1} X_n$.*

Definition 4.4 *Let T be a theory in L and let κ be an infinite cardinal. We say that T is κ-categorical if any two models of T of cardinality κ are isomorphic.*

The theory of algebraically closed fields of characteristic p (for any $p \geq 0$) is κ-categorical for all uncountable κ but has countably many non isomorphic countable models. The theory of dense linear orderings without endpoints is $\aleph_0$-categorical, but for any uncountable κ has 2^κ non isomorphic models of cardinality κ (this is the maximum number possible).

Now consider the space $S_n(\emptyset)$ of complete n-types over the empty set. We know by classical topology that if $S_n(\emptyset)$ is infinite, then it must contain a non isolated point. We also know that either $S_n(\emptyset)$ is countable, in which case the isolated points are dense or it must have cardinality continuum.

Theorem 4.5 (The Ryll-Nardzewski theorem) *Let L be a countable language, let T be a complete theory in L. Then T is $\aleph_0$-categorical iff $S_n(\emptyset)$ is finite for every $n \geq 1$.*

Sketch of proof : Note first that if $S_n(\emptyset)$ is uncountable then T must have uncountably many non isomorphic countable models: any countable model can only realize countably many different types of n-tuples.

Suppose that T is $\aleph_0$-categorical. Then certainly, for every n, $S_n(\emptyset)$ must be countable, hence the isolated types are dense. So for each n, the set X_n of non isolated points in $S_n(\emptyset)$ is meager. The Omitting types theorem says that there is a countable model $\mathcal{M}_0$ of T which realizes only isolated n-types for every n. If for some n $S_n(\emptyset)$ is infinite, then it contains a non isolated type q. This type must be realized in some countable model $\mathcal{M}_1$. Clearly these two models cannot be isomorphic.

Conversely if $S_n(\emptyset)$ is finite for every n, then for all n, all complete n-types over $\emptyset$ are isolated and it is not very difficult to show that any two countable models realizing only isolated types must be isomorphic. □

We mentioned before that it was impossible for all theories to have saturated models in all cardinalities. Here is a first result for countable models. Clearly any countable saturated model must realize every n-type over the empty set. This is only possible if there are only countably many such types. But in fact this is also a sufficient condition.

Theorem 4.6 *Let L be a countable language, let T be a complete theory in L. Then T has a countable saturated model if and only if $S_n(\emptyset)$ is countable for every $n \geq 1$.*

More generally it turns out that the cardinalities of the spaces $S_n(A)$ are quite relevant to the questions we are interested in, related to these ranks we want to define. This will be apparent in the next chapter, but just to give an idea, let us consider briefly the case of theories such that, for all countable A, $S_n(A)$ is countable. This is in particular the case of algrebraically closed fields and differentially closed fields. Theories with this property are called ω-stable.

We can then consider the classical Cantor-Bendixson derivation on the space $S_n(A)$. Recall that if F is compact, one defines the derivative F' of F to be the closed subset of F (which might be empty) obtained by taking away all isolated points of F. One continues the process, taking the derivative of F' etc. More generally define $F^{\alpha+1}$ to be $(F^{\alpha})'$ and take intersections at limit steps. The density of isolated points ensures that this process stops and we can associate to every closed set F (and in particular to every definable set) its Cantor-Bendixson rank, the first α such that F^{α} is empty. This will be a countable ordinal. We cannot use this rank directly but the Morley rank we will define is very similar.

In the particular case of a countable algebraically closed field K of infinite transcendence degree, if we consider the Zariski closed subsets of K^n, then the Cantor-Bendixson rank is equal to the Morley rank and is exactly the algebraic dimension, hence is always finite.

In the case of differentially closed fields, the rank of some of the definable sets is an infinite ordinal.

Separably closed fields are not ω-stable and there is no reasonable way to associate to every definable set such a rank. But another well-behaved rank, the U-rank (or Lascar rank), can be defined on certain families of definable sets. This will be explained in the chapter "Separably closed fields", though not in full detail.

This brings us to a final remark about the topology we are working with in the spaces $S_n(A)$. In the case of fields, this topology is finer than the Zariski topology (where instead of all definable sets, the closed sets are chosen to be the quantifier free "positively" defined ones, i.e. the ones who can be defined by a formula involving no negation and no quantifiers). The question was open for a long time whether it was possible to characterize abstractly the structures on which there exists such a "Zariski-like" topology, necessarily coarser than the one described above. This question was answered recently by Hrushovski and Zilber for the case of structures of "dimension one" and their results, which are at the heart of Hrushovski's proof of the Mordell-Lang conjecture, are presented in the chapter "Zariski geometries" ([Mar]).

References

[ChKe] C. Chang and J. Keisler, *Model Theory*, North-Holland, Amsterdam 1973.

[De] F. Delon, *Separably closed fields*, this volume.

[Ho] W. Hodges, *Model Theory*, Cambridge, 1993.

[Mar 96] D. Marker, *Introduction to the Model Theory of Fields*, in Model Theory of Fields, Lecture Notes in Logic 5, Springer, 1996.

[Mar] D. Marker, *Zariski geometries*, this volume

[Pil] A. Pillay, *Model theory of algebraically closed fields*, this volume.

[Po 85] B. Poizat, *Cours de théorie des modèles*, Nur al-matiq wal ma'rifah, Villeurbanne, France, 1985.

[Wo] C. Wood, *Differentially closed fields*, this volume.

[Zie] M. Ziegler, *Introduction to Stability theory and Morley rank*, this volume.

Introduction to stability theory and Morley rank

Martin Ziegler

We assume knowledge of the basic definitions and results of model theory which are presented (informally) in the first chapter of this book. We use as basic example the theory of algebraically closed fields (see the chapter on algebraically closed fields [Pi1]).

The theory developped here, that of ω-stable theories and Morley rank, is meant to be self contained but it might help to have a look at [Las 87] or [Pi2 96].

Notation and conventions:
We mainly follow the notation introduced in the first chapter except that we now make no distinction in our notation between L-structures and their base sets (i.e. between $\mathcal{M}$ and M).
We consider a **countable** language L and a **complete** theory T in L with infinite models.
We will use the following notation: if M is a model of T, if $\phi(\bar{x})$ is a formula in L,

$$\phi(M) = \{\overline{m} \in M^n; M \models \phi(\overline{m})\}.$$

We will say that a set S is 0-definable if it is $\emptyset$-definable (or definable with no parameters).
We denote by $S(A)$ the set of all complete types over A, that is $S(A) = \bigcup_{n\geq 1} S_n(A)$. We say that a map f from M^n to M^k is *definable* if its graph is a definable subset of M^{n+k}. Of course all the basic maps in the language L are definable but for example in the language L_{ring} for algebraically closed fields of characteristic $p > 0$, both the frobenius map $x \mapsto x^p$ and its inverse $x \mapsto x^{-p}$ are definable maps.

1 The monster model, imaginary elements

Recall that a complete theory T has for arbitrarily large cardinals κ a model N which is *κ-saturated.* This means that N realizes every type over every subset A of N of cardinality strictly less than κ.

Saturated models (κ-saturated models of cardinality κ) of a complete theory are uniquely determined by their cardinality. But they may not exist in the absence of a set theoretical hypothesis, like the general continuum hypothesis for example.

Furthermore it is convenient to work inside a big canonical model which embeds all models. This will be the *monster* model of T.

The universe of the *monster model* is not a set but a proper class (thus it is not a structure in the usual sense) where all types over all sub*sets* are realized. In naive set theory the monster model $\mathfrak{C}$ exists uniquely. It can equivalently be characterized by

1. Every model is elementarily embeddable into $\mathfrak{C}$.

2. Every partial elementary isomorphism between two subsets can be extended to an automorphism of $\mathfrak{C}$.

From now on every set A of parameters is understood to be a subset of $\mathfrak{C}$. A *model* M is now a subset of $\mathfrak{C}$ which is the universe of an elementary substructure. This means that every $L(M)$–formula $\phi(x)$ which is satisfied in $\mathfrak{C}$ can be satisfied by an element of M ([1]). Formulas will have parameters from $\mathfrak{C}$ and we write $\models \phi(c)$ for $\mathfrak{C} \models \phi(c)$. Formulas will define subclasses $\mathbb{D} = \phi(\mathfrak{C})$ rather than subsets.

A note for the cautious: the set theory used for the existence and uniqueness of $\mathfrak{C}$ may be too naive. Thus to keep mathematics inside ZFC one should convince oneself that proofs which make use of the monster model can always be turned into more harmless proofs. The situation is the same in algebra, where category theory is used. We will later assume that T is ω–stable. This implies easily that T has a saturated model N in each cardinality. These models can serve as monster models if they are big enough.

Lemma 1.1 *A definable class $\mathbb{D}$ is definable over A iff $\mathbb{D}$ is invariant under all automorphisms of $\mathfrak{C}$ which leave every element of A fixed. (We call them automorphisms* over *A.)*

It follows that the *definable closure* dcl(A) of A, the set of elements which are definable over A ([2]), is the set of elements which are fixed under every automorphism over A.

An element b contained in a finite A–definable set is *algebraic* over A. Lemma 1.1 implies that b is algebraic over A iff it has only finitely many conjugates over A. We call acl(A), the set of elements algebraic over A, the *algebraic closure* of A

Our **Main Example** will be the theory $\mathrm{T_{ACF}}$ of algebraically closed fields of some fixed characteristic, (see [Pil]). Let A be a subset of the monster field $\mathfrak{C}$. Then dcl(A) is the perfect closure of the subfield K generated by A (i.e. K itself if the characteristic is 0 or $K^{p^{-\infty}}$ if the characteristic is p) and acl(A) is the algebraic closure of K in $\mathfrak{C}$ (in the usual sense).

[1] This is Tarski-Vaught's test.

[2] b is definable over A if the singleton set $\{b\}$ is definable over A.

From $\mathfrak{C}$ we construct the L^{eq}–structure $\mathfrak{C}^{\text{eq}}$ as follows: for every 0-definable equivalence relation

$$E(x_1,\ldots,x_n,y_1,\ldots,y_n)$$

on $\mathfrak{C}^n$ we add the new sort $\mathfrak{C}^n/E$ to $\mathfrak{C}$ and a new function symbol π_E to L for the projection $\mathfrak{C}^n \to \mathfrak{C}^n/E$.

By T^{eq} we denote the theory of $\mathfrak{C}^{\text{eq}}$, $\text{dcl}^{\text{eq}}(A)$ and $\text{acl}^{\text{eq}}(A)$ denote respectively the definable and algebraic closures of A in $\mathfrak{C}^{\text{eq}}$.

The new elements of $\mathfrak{C}^{\text{eq}}$ are called *imaginary* elements. They are used for two main reasons:

1. We can build quotient structures in $\mathfrak{C}^{\text{eq}}$: if for example $\mathbb{G}$ is a definable group in $\mathfrak{C}$ and $\mathbb{H}$ a 0–definable normal subgroup, the quotient group $\mathbb{G}/\mathbb{H}$ can be considered as a group definable in $\mathfrak{C}^{\text{eq}}$.
2. We are concerned about the parameters from which a class can be defined. If for example $\mathbb{F}:\mathbb{G}\to\mathbb{K}$ is a 0-definable homomorphism between two groups in $\mathfrak{C}$, $g\in\mathbb{G}$, the element $k=\mathbb{F}(g)$ is definable from g, but also from any g' in $g\cdot\ker(\mathbb{F})$. It will be more canonical to consider that k is definable from the imaginary element $g\cdot\ker(\mathbb{F})$.

The next lemma shows that we can switch to $\mathfrak{C}^{\text{eq}}$ for most purposes: since $\mathfrak{C}^{\text{eq}} = \text{dcl}^{\text{eq}}(\mathfrak{C})$ every L^{eq}–formula can be translated back to L.

Lemma 1.2 *For every L^{eq}–formula $\phi(x_1,\ldots,x_n)$, where the variable x_i belongs to the sort $\mathfrak{C}^{n_i}/E_i$, there is an L–formula*

$$\psi(\overline{y}_1,\ldots,\overline{y}_n)$$

which, in T^{eq}, is equivalent to

$$\phi(\pi_{E_1}(\overline{y}_1),\ldots,\pi_{E_n}(\overline{y}_n)).$$

Let $\mathbb{D}=\phi(\mathfrak{C},\overline{a})$ be a definable class. Define the equivalence relation $E(\overline{x},\overline{y})$ by

$$E(\overline{b},\overline{c}) \iff \phi(\mathfrak{C},\overline{b})=\phi(\mathfrak{C},\overline{c}).$$

The element $c=\overline{a}/E$ of $\mathfrak{C}^{\text{eq}}$ is a *canonical parameter* of $\mathbb{D}$ in the sense of the following definition:

Definition 1.3 *A tuple $\overline{c}$ is a* canonical parameter *of the definable class $\mathbb{D}$ if it is fixed pointwise by exactly those automorphisms which fix $\mathbb{D}$ setwise.*

By 1.1 $\overline{c}$ is determined up to interdefinability and in general it will exist only in $\mathfrak{C}^{\text{eq}}$, but we give it a name: $\text{cb}(\mathbb{D})$.

<u>A trivial example</u>
If $\mathbb{D}$ is a coset of a 0–definable equivalence relation on $\mathfrak{C}^n$ a canonical parameter of $\mathbb{D}$ is $\mathbb{D}$ itself considered as an element of $\mathfrak{C}^{\text{eq}}$.

An n–tuple $\overline{a}$ of elements of $\mathfrak{C}$ can thus be identified with the singleton set $\{\overline{a}\}$, considered as its own canonical parameter. *This will allow us, when working in $\mathfrak{C}^{\text{eq}}$, to simplify our notation and to make no difference between tuples and elements unless we have a specific reason to do it.*

A theory has *elimination of imaginaries* if every definable class has a canonical parameter (in $\mathfrak{C}$). An easy extension of the above construction yields

Lemma 1.4 *T^{eq} has elimination of imaginaries.*

The theory of an infinite set without structure does not eliminate imaginaries since two–element sets do not have a canonical parameter.

The theory T_{ACF} eliminates imaginaries (see [Pi1]). This is a slight generalization of a theorem of Weil, which says that every algebraic variety has a smallest field of definition. The two–element set $\{a, b\}$ for example has the pair $(a + b, ab)$ as a canonical parameter.
Weil's theorem also extends to differentially closed fields and separably closed fields of finite degree of imperfection, which eliminate imaginaries (see [Wo] and [De]).

Lemma 1.5 *Suppose that in T, there are two distinct 0–definable elements. Then the following are equivalent*

1. *T eliminates imaginaries*
2. *Every element of $\mathfrak{C}^{\text{eq}}$ is interdefinable with an n–tuple from $\mathfrak{C}$.*
3. *Any 0–definable class in $\mathfrak{C}^{\text{eq}}$ is in 0-definable bijection with a 0–definable subclass of $\mathfrak{C}^n$.*

Proof: If c is a canonical parameter of $\mathbb{D}$ the other canonical parameters are the tuples interdefinable with c. This proves $1\leftrightarrow 2$.
If $f : \mathfrak{C}^m/E \to \mathfrak{C}^n$ is a 0–definable injection c and $f(c)$ are interdefinable; $3\to 2$ follows.
For $2\to 3$ we note first that $c \in \mathfrak{C}^m/E$ and $a \in \mathfrak{C}^n$ are interdefinable iff $f(c) = a$ for a 0–definable injective function, defined on a subset of $\mathfrak{C}^m/E$. If every c is interdefinable with some tuple a a compactness argument shows that all the f's can be taken from a finite set $\{f_1, \ldots, f_k\}$ of functions. Our assumption that we have two distinct constants gives us k different 0–definable elements in some $\mathfrak{C}^n$. We can then patch the f_i's together to get a 0–definable injection from $\mathfrak{C}^m/E$ to some $\mathfrak{C}^k$. □

Let c be a canonical parameter of $\mathbb{D}$ and A a set of parameters. It is easy to see that $\mathbb{D}$ is definable over A iff $c \in \text{dcl}(A)$.

Lemma 1.6 *The following are equivalent:*

1. $c \in \text{acl}(A)$

2. *$\mathbb{D}$ is definable over $acl(A)$*
3. *$\mathbb{D}$ has only finitely many conjugates over A*
4. *$\mathbb{D}$ consists of classes of a* finite *A-definable equivalence relation.*

(A *finite equivalence relation* is an equivalence relation with finitely many classes.)
Proof: 1↔2: clear
2↔3: $\mathbb{D}$ has as many conjugates as c.
3→4: let $\mathbb{D}_0, \ldots, \mathbb{D}_{k-1}$ be the A-conjugates of $\mathbb{D}$. Then

$$\bigwedge_{i<k} (x \in \mathbb{D}_i \leftrightarrow y \in \mathbb{D}_i)$$

defines an equivalence relation with at most 2^k many classes, which is invariant under automorphisms over A.
4→3: clear □

Definition 1.7 *The strong type* $\mathrm{st}(a/A)$ *of a over A is the type of a over* $\mathrm{acl}^{\mathrm{eq}}(A)$.

Corollary 1.8 $\mathrm{st}(a/A)$ *can be axiomatized by*

$$\{E(x,a) \mid E \text{ an } A\text{-definable finite equivalence relation}\}.$$

2 Morley rank

Let T be as in section 1 and let $\mathfrak{C}$ be its monster model.

We want to attach to each definable class $\mathbb{D}$ an ordinal number (or -1 and ∞), its *Morley rank*, denoted *MR*. First we define the relation $\mathrm{MR}(\mathbb{D}) \geq \alpha$ by recursion on the ordinal α:

Definition 2.1

$\mathrm{MR}(\mathbb{D}) \geq 0$	*iff*	*$\mathbb{D}$ is not empty*
$\mathrm{MR}(\mathbb{D}) \geq \lambda$	*iff*	*$\mathrm{MR}(\mathbb{D}) \geq \alpha$ for all $\alpha < \lambda$, (λ a limit ordinal)*
$\mathrm{MR}(\mathbb{D}) \geq (\alpha+1)$	*iff*	*there is an infinite family $(\mathbb{D}_i)$ of disjoint definable subclasses of $\mathbb{D}$ such that $\mathrm{MR}(\mathbb{D}_i) \geq \alpha$ for all i.*

Then the Morley rank of $\mathbb{D}$ is

$$\mathrm{MR}(\mathbb{D}) = \sup\{\alpha \mid \mathrm{MR}(\mathbb{D}) \geq \alpha\},$$

with the convention that $\mathrm{MR}(\emptyset) = -1$ and $\mathrm{MR}(\mathbb{D}) = \infty$ if $\mathrm{MR}(\mathbb{D}) \geq \alpha$ for all α (here we say that $\mathbb{D}$ has *no rank*). Note that the meaning of $\mathrm{MR}(\mathbb{D}) \geq \alpha$ has not changed.

Special cases. a definable class has rank -1 iff it is empty, rank 0 iff it is finite and not empty and rank 1 iff it is infinite but does not contain an infinite family of disjoint infinite definable classes.

Lemma 2.2

$$\mathrm{MR}(\mathbb{D}_1 \cup \mathbb{D}_2) = \max(\mathrm{MR}(\mathbb{D}_1), \mathrm{MR}(\mathbb{D}_2))$$

Proof: The inequality $\mathrm{MR}(\mathbb{D}_1 \cup \mathbb{D}_2) \geq \max(\mathrm{MR}(\mathbb{D}_1), \mathrm{MR}(\mathbb{D}_2))$ is clear. The other follows from

$$\mathrm{MR}(\mathbb{D}_1 \cup \mathbb{D}_2) \geq \alpha \Rightarrow \mathrm{MR}(\mathbb{D}_1) \geq \alpha \text{ or } \mathrm{MR}(\mathbb{D}_2) \geq \alpha,$$

which is easily proved by induction on α. □

The lemma implies that the classes of smaller rank than α form an ideal in the boolean algebra of all definable classes. We fix notation for equality and inclusion in the quotient algebra:

$$\mathbb{F} \subset_\alpha \mathbb{G} \quad \text{if} \quad \mathrm{MR}(\mathbb{F} \setminus \mathbb{G}) < \alpha$$
$$\mathbb{F} =_\alpha \mathbb{G} \quad \text{if} \quad \mathrm{MR}(\mathbb{F} \Delta \mathbb{G}) < \alpha$$

A definable class $\mathbb{F}$ of rank α cannot contain an infinite family (G_i) of (mod α)-disjoint subclasses of rank α. (Otherwise one could replace the G_i by (mod α)–equivalent sets G'_i which are really disjoint.) This means that in the quotient algebra $\mathbb{F}$ consists of finitely many atoms. The number of these atoms is the *Morley degree* of $\mathbb{F}$.

Definition 2.3 *The* Morley degree $\mathrm{Md}(\mathbb{D})$ *of a class* $\mathbb{D}$ *of Morley rank* α *is the maximal length* d *of a decomposition* $\mathbb{D} = \mathbb{D}_1 \dot\cup \ldots \dot\cup \mathbb{D}_d$ *into classes of rank* α.

The components $\mathbb{D}_i$ of $\mathbb{D}$ are uniquely determined mod α.

In the rank 0 case the degree of $\mathbb{D}$ is simply the number of elements of $\mathbb{D}$. If $\mathbb{D}$ has no rank the Morley degree of $\mathbb{D}$ is not defined.

Definition 2.4 *A* strongly minimal *class is a definable class of rank one and degree one.*

$\mathbb{D}$ is strongly minimal iff it is infinite and every definable subclass is finite or cofinite in $\mathbb{D}$. Algebraically closed fields are strongly minimal since non trivial equations and inequalities define finite and cofinite sets.

Lemma 2.5 *Assume that* $\mathbb{F}$ *and* $\mathbb{G}$ *are disjoint and that* $\mathrm{MR}(\mathbb{F}) \leq \mathrm{MR}(\mathbb{G}) < \infty$. *Then*

$$\begin{aligned} \mathrm{Md}(\mathbb{F} \cup \mathbb{G}) &= \mathrm{Md}(\mathbb{F}) + \mathrm{Md}(\mathbb{G}) && \text{, if } \mathbb{F} \text{ and } \mathbb{G} \text{ have the same rank} \\ \mathrm{Md}(\mathbb{F} \cup \mathbb{G}) &= \mathrm{Md}(\mathbb{G}) && \text{, otherwise.} \end{aligned}$$

Let p be a type over A. If p does not contain a ranked formula the *Morley rank* $\mathrm{MR}(p)$ of p is ∞ by definition. Otherwise we choose a formula ϕ in p which has minimal rank α and, among the formulas of rank α, has minimal Morley degree d. We define $\mathrm{MR}(p)$ to be α and the *Morley degree* $\mathrm{Md}(p)$ of p as d. A type of rank and degree 1 is called *strongly minimal*

It is easy to see that ϕ is – up to equivalence mod α – uniquely determined by p. Conversely ϕ *determines* p as the only type over A which contains ϕ and has at least the rank α. In fact one can recover p as

$$p = \{\psi(x) \text{ in } L(A) \mid \phi \subset_\alpha \psi\}.$$

The $L(A)$–formulas which belong in the way described above to ranked types over A are exactly those formulas which cannot be decomposed into two A–definable subclasses of the same rank. From this one checks easily that for $L(A)$–formulas $\phi(x)$

$$\mathrm{MR}(\phi) = \max\{\mathrm{MR}(p) \mid p \in \mathrm{S}(A),\ \phi \in p\} \tag{2.1}$$

$$\mathrm{Md}(\phi) = \sum \mathrm{Md}(p) \text{ for } p \in \mathrm{S}(A),\ \phi \in p,\ \mathrm{MR}(p) = \mathrm{MR}(\phi). \tag{2.2}$$

Let $\mathbb{D}$ be a class of rank α which is defined over a parameter set A. We call an element a of $\mathbb{D}$ *generic* (over A) if $\mathrm{MR}(a/A) = \alpha$. (We use the notation $\mathrm{MR}(a/A)$ for $\mathrm{MR}(\mathrm{t}(a/A))$.) If $\phi(x)$ is as above the type $\mathrm{t}(a/A)$ of a generic element of $\phi(\mathfrak{C})$ is the (unique) type determined by ϕ over A.

Definition 2.6

1. *T is* totally transcendental *if every definable class has Morley rank.*
2. *T is* ω–stable *if* $\mathrm{S}_1(A)$ *is countable for every countable A.*

The theory of dense linear orders without first and last elements for example is not ω–stable: distinct Dedekind cuts define distinct types over the rationals. Algebraically closed fields are ω–stable: the types over a subfield K are the type of a transcendental element and the algebraic types given by irreducible polynomials over K. Hence $|\mathrm{S}_1(K)| = |K| + \omega$.

Theorem 2.7 *T is totally transcendental iff T is ω–stable.*

The "only if" direction uses the countability of L.
Proof: If T is totally transcendental every type over A is determined by an $L(A)$–formula. Whence $|\mathrm{S}_1(A)| \leq |L(A)\text{–formulas}| = |A| + \omega$.
For the converse we observe first that conjugate definable classes have the same rank. Therefore there are at most as many ranks as there are types of parameter tuples over $\emptyset$ and we can conclude that the class of possible ranks is a set. (In fact it is a countable ordinal!). It follows now from the definition that every definable class without rank contains two disjoint classes without rank. If T is not totally transcendental, starting from $\mathfrak{C}_\emptyset = \mathfrak{C}$, we can construct a binary tree $(\mathfrak{C}_s)_{s \in {}^{<\omega}2}$ of definable classes without rank such that

1. $\mathfrak{C}_s \neq \emptyset$
2. $\mathfrak{C}_{si} \subset \mathfrak{C}_s$ $(i < 2)$
3. $\mathfrak{C}_{s0} \cap \mathfrak{C}_{s1} = \emptyset$.

Choose a countable set A which contains parameters for the $\mathfrak{C}_s$ and choose for every path $\sigma \in {}^{\omega}2$ a type $p_\sigma \in \mathrm{S}_1(A)$ which contains all formulas $\mathfrak{C}_s(x)$ ($s \subset \sigma$). The p_σ are all different, which shows that $\mathrm{S}_1(A)$ has continuum many elements. □

The Morley rank for definable subclasses of $\mathfrak{C}^n$ is defined in the same way as for $\mathfrak{C}$.

Corollary 2.8 *If T is ω–stable, then every definable subclass of $\mathfrak{C}^n$ is ranked. In fact T^{eq} is ω–stable.*

Proof: Assume that T is ω–stable. Then $\mathrm{S}_n(A)$ is countable for all countable A and all integers n. We prove this by induction on n: let A be a countable set of parameters. By induction there is a countable set B where every $(n-1)$–type over A is realized and again a countable set C where every 1–type over B is realized. It is now easy to see that every n–type over A is realized in C. This shows that $\mathrm{S}_n(A)$ is countable. That T^{eq} is ω–stable follows from the fact that the projections $\mathfrak{C}^n \to \mathfrak{C}^n/E$ define surjections $\mathrm{S}_n(A) \to \mathrm{S}'(A)$, where S' means 1–types in a variable for the sort $\mathfrak{C}^n/E$. □

A closer look at the proof shows that the cartesian product of $\mathbb{D}_1, \ldots, \mathbb{D}_n$ is ranked whenever each $\mathbb{D}_i$ is ranked, which can be reformulated as: if the types $\mathrm{t}(a_1/A), \ldots, \mathrm{t}(a_n/A)$ are ranked then $\mathrm{t}(a_1 \ldots a_n/A)$ also has a rank. In fact a bound for the rank of $\mathrm{t}(a_1 \ldots a_n/A)$ can be computed from the ranks of the $\mathrm{t}(a_1/A), \ldots, \mathrm{t}(a_n/A)$.

Lemma 2.9 *If b is algebraic over Aa then* $\mathrm{MR}(b/A) \leq \mathrm{MR}(a/A)$.

Proof: To simplify notation assume that A is empty (just add A as constants in L). We prove

$$\mathrm{MR}(b) \geq \alpha \Rightarrow \mathrm{MR}(a) \geq \alpha$$

by induction on α.
If α is 0 or a limit ordinal there is nothing to show. So assume $\mathrm{MR}(b) \geq \alpha = \beta+1$ and $\mathrm{MR}(a) = \beta$ for contradiction. Choose a 0–definable class $\mathbb{D}$ of rank β which contains a. Let the algebraicity of b over a be witnessed by an L–formula $\phi(x, y)$ such that $\models \phi(a, b)$ and $\phi(a', \mathfrak{C})$ has at most (say) k elements for all a'. Since the class $\mathbb{E}$ defined by $\mathbb{E}(y) = \exists x\, \mathbb{D}(x) \wedge \phi(x, y)$ contains b it has a rank greater than β and we find an infinite family $(\mathbb{E}_i)$ of disjoint definable subclasses of $\mathbb{E}$ of rank $\geq \beta$. For each $\mathbb{E}_i$ define $\mathbb{D}_i$ by $\mathbb{D}_i(x) = \exists y\, \mathbb{D}(x) \wedge \phi(x, y) \wedge \mathbb{E}_i(y)$. Let $\mathbb{E}_i$ be definable over the parameters e'. Choose $b' \in \mathbb{E}_i$ of rank $\geq \beta$ over e' and $a' \in \mathbb{D}_i$ with $\models \phi(a', b')$. The induction hypothesis implies that a' has also rank $\geq \beta$ over e'. Whence the rank of $\mathbb{D}_i$ is at least β and $\mathbb{D}_i$ contains (mod β) one of the finitely many components of $\mathbb{D}$. This implies that there is an infinite subfamily of $(\mathbb{D}_i)$ with non–empty intersection. But an element of such an intersection would be ϕ–related to infinitely many elements, which is impossible. □

Corollary 2.10 *If there is a definable bijection $\mathbb{F} : \mathbb{D} \to \mathbb{E}$ the two classes $\mathbb{D}$ and $\mathbb{E}$ have the same rank and degree.*

From now on we will use the remark made earlier that n–tuples can be considered as single elements in $\mathfrak{C}^{eq}$ and will in general make no difference in our notation between n-tuples and elements.

Lemma 2.11 *Let M be ω-saturated. Then the rank of an M-definable class can be computed using only M-definable classes, which means that the rank of an M-definable class $\mathbb{D}$ is bigger than α iff there is an infinite family of M-definable disjoint subclasses of $\mathbb{D}$ of rank $\geq \alpha$.*

Proof: An M–definable class $\phi(\mathfrak{C}, m)$ of Morley rank $\geq (\alpha + 1)$ contains an infinite family

$$\phi_0(\mathfrak{C}, b_0),\ \phi_1(\mathfrak{C}, b_1), \ldots$$

of disjoint definable classes which have a rank at least α. Find a sequence $m_0, m_1, \ldots$ in M such that the (whole) sequence has the same type over m as $b_0, b_1, \ldots$. Then

$$\phi_0(\mathfrak{C}, m_0),\ \phi_1(\mathfrak{C}, m_1), \ldots$$

is an infinite sequence of disjoint subclasses of $\phi(\mathfrak{C}, m)$ of rank at least α, which are definable over M. □

The same proof shows

Corollary 2.12 *Let M be an ω-saturated model and ϕ an $L(M)$-formula of rank α and Morley degree d. Then ϕ can be decomposed into $L(M)$-formulas $\phi_1, \ldots, \phi_d$ of rank α and degree 1.*

It will follow from 4.10 that for ω–stable theories the ω–saturation of M is not needed (since types over models have degree 1).

3 Definable types

Lemma 3.1 *Let M be an ω-saturated model. If $\mathbb{E}$ is non-empty, has a rank and is contained in an M-definable class of the same rank then $\mathbb{E}$ intersects M.*

We will see later (4.14) that for ω-stable theories "non–forking" extensions of types over models are "coheirs". This implies that for ω–stable theories the above is true for arbitrary models.

Proof: Let $\mathbb{D}$ be an M–definable class which contains $\mathbb{E}$ and has the same rank α. We prove the lemma by induction on α and the degree d of $\mathbb{D}$.

If $\alpha = 0$ $\mathbb{D}$ is finite and contained in M.

If $d > 1$ we decompose $\mathbb{D} = \mathbb{D}_1 \cup \ldots \cup \mathbb{D}_d$ into its components of degree 1. By 2.12 we can assume that the $\mathbb{D}_i$ are over M. $\mathbb{E}$ must intersect one component $\mathbb{D}_i$ in a class of rank α and we can apply the induction hypothesis to $\mathbb{D}_i \cap \mathbb{E}$ inside $\mathbb{D}_i$.

Finally assume that $\alpha > 0$ and $d = 1$. Then $\mathbb{D} \setminus \mathbb{E}$ has a rank β smaller then α. This implies that $\mathbb{D}$ contains an infinite family $(\mathbb{D}_i)$ of M–definable subclasses of rank $\geq \beta$ but smaller than α. (2.11). But $\mathbb{D} \setminus \mathbb{E}$ cannot contain such a family. Therefore the rank of some $(\mathbb{D} \setminus \mathbb{E}) \cap \mathbb{D}_i$ must be smaller than β, and we can apply the induction hypothesis to $\mathbb{E} \cap \mathbb{D}_i$. □

Theorem 3.2 *Let $\mathbb{D}$ be a definable class of rank α and $\phi(x,y)$ an L-formula. Then*

$$\{b \mid \mathbb{D} \subset_\alpha \phi(\mathfrak{C}, b)\}$$

is definable.

For the proof we introduce the following classical definitions:

Definition 3.3

1. *A formula $\phi(\overline{x}, \overline{y})$ has the* order property *if there are tuples $\overline{a}_i$, $\overline{b}_i$ $(i < \omega)$ such that*

$$\models \phi(\overline{a}_i, \overline{b}_j) \iff i \leq j.$$

2. *T is* stable *if there is no formula with the order property.*

Lemma 3.4 *ω-stable theories are stable.*

Proof: Assume that $\phi(x,y)$ has the order property witnessed by a_i, b_i. A compactness argument shows that we can replace the index set ω by the order $\mathbb{Q}$ of the rationals. Now choose $\mathbb{D}$ of minimal rank and degree such that $I = \{i \in \mathbb{Q} \mid a_i \in \mathbb{D}\}$ is an interval of positive length. Fix j in the interior of I. Then both $D(x) \wedge \phi(x, b_j)$ and $D(x) \wedge \neg\phi(x, b_j)$ define intervals of positive length. But one of the two formulas must have smaller rank or smaller degree than $\mathbb{D}$. □

Proof of 3.2:
Since it is enough to prove the theorem for each of the degree 1 components of $\mathbb{D}$ we can assume that $\mathbb{D}$ has degree 1.
Fix an ω-saturated model M which contains parameters for $\mathbb{D}$.
Claim 1: $\mathbb{D}(M)$ contains a finite set Δ such that for all b

$$\Delta \subset \phi(\mathfrak{C}, b) \implies \mathbb{D} \subset_\alpha \phi(\mathfrak{C}, b). \tag{2.1}$$

Proof: If Δ does not exist we construct a sequence (a_i, b_i) as follows: assume that $a_0, b_0, \ldots a_{n-1}, b_{n-1}$ are already constructed and such that the a_i's are elements of $\mathbb{D}(M)$ and the $\phi(\mathfrak{C}, b_i)$ do not contain $\mathbb{D}$ (mod α). Since $\mathbb{D}$ has degree 1 we have $\mathbb{D} \not\subset_\alpha \phi(\mathfrak{C}, b_0) \cup \ldots \cup \phi(\mathfrak{C}, b_{n-1})$. By 3.1 there is $a_n \in M$ which belongs to $\mathbb{D}$ but not to $\phi(\mathfrak{C}, b_0) \cup \ldots \cup \phi(\mathfrak{C}, b_{n-1})$. Since $\{a_0, \ldots, a_n\}$ does not serve as Δ there is a b_n such that $\{a_0, \ldots, a_n\} \subset \phi(\mathfrak{C}, b_n)$ and $\mathbb{D} \not\subset_\alpha \phi(\mathfrak{C}, b_n)$. The construction shows that the sequence (a_i, b_i) is ordered by ϕ as in 3.4. But by the proof there $\mathbb{D}$ cannot be ranked.
Claim 2: For each c such that $\mathbb{D} \subset_\alpha \phi(\mathfrak{C}, c)$ there is a finite $\Delta \subset \phi(\mathfrak{C}, c)$ with satisfies (2.1). Δ can be chosen as a subset of $\mathbb{D}(M)$.
Proof: The a_i in the the proof of Claim 1 can be found in $\phi(\mathfrak{C}, c)$.

Let $\mathfrak{D}$ be the set(!) of all $\Delta \subset \mathbb{D}(M)$ which satisfy (2.1) and let $\chi_\Delta(y)$ be the formula which says "$\Delta \subset \phi(\mathfrak{C}, y)$". Then by claim 2 "$\mathbb{D} \subset_\alpha \phi(\mathfrak{C}, y)$" is definable by the infinite disjunction $\bigvee_{\Delta \in \mathfrak{D}} \chi_\Delta(y)$. Since $\mathbb{D}$ has degree 1 the negation "$\mathbb{D} \not\subset_\alpha \phi(\mathfrak{C}, y)$" is equivalent to "$\mathbb{D} \subset_\alpha (\neg\phi)(\mathfrak{C}, y)$" and is therefore also definable by an infinite disjunction. By compactness there must be a finite subset $\mathfrak{D}_0$ of $\mathfrak{D}$ such that for all b

$$\mathbb{D} \subset_\alpha \phi(\mathfrak{C}, b) \Longleftrightarrow \bigvee_{\Delta \in \mathfrak{D}_0} \chi_\Delta(b). \qquad \square$$

Lemma 1.1 shows that $\{b \mid \mathbb{D} \subset_\alpha \phi(\mathfrak{C}, b)\}$ is definable using the parameters of $\mathbb{D}(x)$ only. On the other hand the proof of 3.2 shows that one can define $\{b \mid \mathbb{D} \subset_\alpha \phi(\mathfrak{C}, b)\}$ over $\mathbb{D}(M)$ for any ω–saturated model M over which $\mathbb{D}$ can be defined or, more generally, over any set in which every rank α subclass of $\mathbb{D}$ is realized.

Corollary 3.5 *(T ω–stable)*
Every type $p \in S(A)$ is definable *over A in the following sense: for each L–formula $\phi(x, \overline{y})$ there is an $L(A)$–formula $d_p x\phi(x, \overline{y})$ such that for all $\overline{a}$ from A*

$$\phi(x, \overline{a}) \in p \Longleftrightarrow \models d_p x \phi(x, \overline{a}).$$

Proof: Choose a formula $\mathbb{D}(x) \in p$ which determines p in the sense of section 2. $\square$

The corollary applied to types over proper classes shows that every *global* type (i.e. a type $\mathsf{p} \in S(\mathfrak{C})$) has a *canonical base* in the following sense:

Definition 3.6 *A* canonical base *of a global type* p, *denoted by* $\mathrm{cb}(\mathsf{p})$, *is a set which is fixed pointwise by exactly those automorphisms which fix* p.

Take for $\mathrm{cb}(\mathsf{p})$ the union of all canonical parameters for all the definable classes $d_{\mathsf{p}} x\phi(x, \mathfrak{C}^n)$. Note that $\mathrm{cb}(\mathsf{p})$ is only defined up to interdefinability. We will see in 4.12 that $\mathrm{cb}(\mathsf{p})$ can be chosen to be a finite set.

Corollary 3.7 *(T ω–stable)*
Let $\mathbb{D}$ be definable over A. Then every definable subclass of $\mathbb{D}$ is definable with parameters from $A \cup \mathbb{D}$.

Proof: Every subclass $\psi(\mathfrak{C}, c)$ of $\mathbb{D}$ equals $\{a \in \mathbb{D} \mid \psi(a, y) \in \mathrm{t}(c/\mathbb{D})\}$. $\square$

4 Forking

Throughout this section T is assumed to be ω–stable.

Definition 4.1 *Let A be a subset of B and p a type over A. A* nonforking extension *q of p is a type over B which extends p and has the same rank as p.*

If q is a nonforking extension of $q{\restriction}_A$, we say that q *does not fork over A*.

Some special cases: let q be an extension of p as above. Then p is algebraic iff q is algebraic and a nonforking extension of p. If p has rank 1 then q is a nonforking extension of p iff it is non-algebraic.

Lemma 4.2 (Existence) *Every type p over A has a nonforking extension to B. There are at most* $\mathrm{Md}(p)$ *many nonforking extensions. More precisely:*

$$\mathrm{Md}(p) = \sum \mathrm{Md}(q) \quad \textit{for } q \in \mathrm{S}(B)\ ,\ q \textit{ a nonforking extension of } p.$$

Proof: If p has rank α and is determined by the formula ϕ the nonforking extensions of p are those types over B which have rank α and contain ϕ. The lemma follows now from property (2.2) in section 2. □

If B is large enough so that ϕ decomposes into B–definable formulas of rank α and degree 1, p has exactly $\mathrm{Md}(p)$ nonforking extensions. We therefore call $\mathrm{Md}(p)$ the *multiplicity* of p. Types with multiplicity 1 are *stationary* :

Definition 4.3 *A type $p \in \mathrm{S}(A)$ is* stationary *if it has exactly one nonforking extension to every set which contains A. For stationary p the canonical base of p,* $\mathrm{cb}(p)$, *is the canonical base of the unique nonforking global extension of p.*

As an example take two non–zero elements a, b of a subfield K of the algebraically closed field $\mathfrak{C}$. The pointed curve $\mathbb{D}(x,y) = (ax^2 \doteq y \wedge \neg x \doteq b)$ has (rank 1 and) degree 1. Whence it determines a stationary type p over K, the global nonforking extension of which can be axiomatized by the formulas $ax^2 \doteq y$ and $\neg x \doteq c$ for all $c \in \mathfrak{C}$. It is now easily seen that $\mathrm{cb}(p) = a$.

Lemma 4.4 (Transitivity and Monotonicity) *Let $p \subset q \subset r$ be a chain of types. Then r is a non–forking extension of p iff r is a nonforking extension of q and q is a nonforking extension of p.*

Lemma 4.5 (Continuity) *Let B be a set of parameters, $q \in \mathrm{S}(B)$ and A a subset of B.*

1. *$q \in \mathrm{S}(B)$ does not fork over some finite subset B_0 of B. Moreover B_0 can be chosen in such a way that q is the unique nonforking extension of $q{\restriction}_{B_0}$ to B.*

2. *If $q \in \mathrm{S}(B)$ forks over A there is a finite subset $B_0 \subset B$ such that $q{\restriction}_{(A \cup B_0)}$ forks over A.*

Proof: Let $\psi \in q$ be of the same rank as q. For B_0 take a set which contains the parameters of ψ. For the "moreover" part ψ should also have the same degree as p. □

Definition 4.6 *We say that a and B are* independent *over A if* $\mathrm{t}(a/AB)$ *does not fork over A.*
We denote it by :

$$a \mathop{\downarrow}_{A} B.$$

Theorem 4.7 (Symmetry) $a \downarrow_A b$ *implies* $b \downarrow_A a$.

Proof: Let us first assume that $A = M$ is ω–saturated. Set $\alpha = \mathrm{MR}(a/M)$ and $\beta = \mathrm{MR}(b/M)$. Let $\mathrm{t}(a/M)$ be determined by $\phi(x)$ and $\mathrm{t}(b/M)$ be determined by $\psi(y)$. If $b \not\perp_M a$ the pair ab satisfies a formula $\chi(x,y)$ over M such that $\mathrm{MR}\chi(a,y) < \beta$. We can assume that $\chi(x,y)$ implies $\phi(x)$ and $\psi(y)$. By 3.2 the class

$$\{c \mid \mathrm{MR}\chi(c,y) < \beta\} = \{c \mid \psi(\mathfrak{C}) \not\subseteq_\beta \chi(c,\mathfrak{C})\}$$

is definable by an $L(M)$–formula $\phi'(x)$. Replacing $\chi(x,y)$ by $\chi(x,y) \wedge \phi'(x)$ allows us to assume that $\models \chi(c,d)$ implies $\mathrm{MR}\chi(c,y) < \beta$ for all c, d. Whence $\chi(x,b)$ cannot be realized in M, which by 3.1 implies that $\mathrm{MR}\chi(x,b) < \mathrm{MR}\phi(x) = \alpha$. This shows $a \not\perp_M b$.

If A is an arbitrary set extend it to an ω–saturated M. Since nonforking extensions exist we can assume that

$$(1) \quad b \mathop{\perp}_{A} M \qquad \text{and} \qquad (2) \quad a \mathop{\perp}_{Ab} M.$$

Using transitivity and monotonicity (i.e. trivial rank inequalities) one sees that (2) implies $a \perp_A b \Rightarrow a \perp_M b$ and (1) implies $b \perp_M a \Rightarrow b \perp_A a$. □ Symmetry

and continuity guarantee that we can extend the notion of independence from finite tuples to arbitrary sets without loosing its basic properties:

Definition 4.8 *B is independent from C over A, denoted $B \perp_A C$, if all finite tuples $\bar{b}$ from B are independent from C over A.*

Corollary 4.9 *A type over* $\mathrm{acl}(A)$ *does not fork over A.*

Theorem 4.10 *Every type over A is stationary provided A is algebraically closed in $\mathfrak{C}^{\mathrm{eq}}$. In other words: strong types are stationary*

Proof: Assume A to be algebraically closed in $\mathfrak{C}^{\mathrm{eq}}$.

We note first that any global type r which does not fork over A is definable over A. This is because r has at most $\mathrm{Md}(\mathrm{r} \restriction A)$ conjugates over A. Therefore the canonical base (as a long tuple) has at most $\mathrm{Md}(\mathrm{r}\restriction_A)$ many A-conjugates and must be contained in $\mathrm{acl}^{\mathrm{eq}}(A) = A$.

Let q_1 and q_2 be two nonforking extensions of p to some set B. We want to show that $q_1 = q_2$ and for this we can assume that $B = Ab$ is an extension of A by a finite tuple b. Realize q_1 and q_2 by two elements a_1 and a_2 which are independent over B. Now by the laws of forking:

$$a_2 \mathop{\perp}_{Ab} a_1 \overset{\text{transitivity}}{\Longrightarrow} a_2 \mathop{\perp}_{A} a_1 b \overset{\text{monotonicity}}{\Longrightarrow} a_2 \mathop{\perp}_{Aa_1} b \overset{\text{symmetry}}{\Longrightarrow} b \mathop{\perp}_{Aa_1} a_2$$

$$\overset{\text{transitivity}}{\Longrightarrow} b \mathop{\perp}_{A} a_1 a_2.$$

Since we can extend the type $\mathrm{t}(b/Aa_1a_2)$ to a global type which does not fork over A it is definable over A. Since a_1 and a_2 have the same type over A we have $\mathrm{t}(ba_1/A) = \mathrm{t}(ba_2/A)$. Whence $q_1 = q_2$. □

It follows that every type over a model M is stationary. M is not algebraically closed in $\mathfrak{C}^{\mathrm{eq}}$ but every type over M extends uniquely to the definable closure M^{eq} of M in $\mathfrak{C}^{\mathrm{eq}}$, which is algebraically closed.

Corollary 4.11 (Conjugacy) *All global nonforking extensions of $p \in \mathrm{S}(A)$ are conjugate over A.*

Proof: If q and r are two global nonforking extensions of p, consider the stationary types $q = \mathsf{q}\restriction_{\mathrm{acl}^{\mathrm{eq}}(A)}$ and $r = \mathsf{r}\restriction_{\mathrm{acl}^{\mathrm{eq}}(A)}$. It is easy to see that all extensions of p to $\mathrm{acl}^{\mathrm{eq}}(A)$ are conjugate over A since $\mathrm{acl}^{\mathrm{eq}}(A)$ is fixed setwise by any A-automorphism. If f is an automorphism of $\mathfrak{C}$ over A which maps q to r, f also maps q to r since q and r are the unique nonforking global extensions of q and r. □

Theorem 4.12 *Let p be a global type and A a set of parameters.*

1. *p does not fork over A iff $\mathrm{cb}(\mathsf{p}) \subset \mathrm{acl}^{\mathrm{eq}}(A)$.*
2. *p is the unique nonforking extension of (the stationary type) $\mathsf{p}\restriction_A$ iff $\mathrm{cb}(\mathsf{p}) \subset \mathrm{dcl}^{\mathrm{eq}}(A)$*

It follows that $\mathrm{cb}(\mathsf{p})$ can be chosen as a finite set.

Proof: 1: That p is definable over $\mathrm{acl}^{\mathrm{eq}}(A)$ if it does not fork over A was already shown in the proof of 4.10.
For the converse assume that p is definable over $\mathrm{acl}^{\mathrm{eq}}(A)$. Let B be any set which contains A. We will show that $\mathsf{p}\restriction_B$ does not fork over A. Choose a model M containing A which is independent from B over A. p and the global nonforking extension of $\mathsf{p} \restriction M$ are both definable over M. Whence the two types must be must be equal by the next lemma. It follows that p does not fork over M and $a \mathop{\downarrow}_M B$ if a realizes $\mathsf{p}\restriction_{BM}$. We have to show that $a \mathop{\downarrow}_A B$. But by the laws of forking:

$$a \mathop{\downarrow}_M B \overset{Symmetry}{\Longrightarrow} B \mathop{\downarrow}_M a \overset{Transitivity}{\Longrightarrow} B \mathop{\downarrow}_A Ma \overset{Monotonicity}{\Longrightarrow} B \mathop{\downarrow}_A a \overset{Symmetry}{\Longrightarrow} a \mathop{\downarrow}_A B.$$

2: Assume that p does not fork over A or equivalently that $\mathrm{cb}(\mathsf{p}) \subset \mathrm{acl}^{\mathrm{eq}}(A)$. By 4.11 p is the unique nonforking extension of $\mathsf{p}\restriction_A$ iff p is invariant under all A–automorphisms of $\mathfrak{C}$ iff the elements of $\mathrm{cb}(\mathsf{p})$ are fixed by all A–automorphisms. This last condition means that $\mathrm{cb}(\mathsf{p}) \subset \mathrm{dcl}^{\mathrm{eq}}(A)$ (by 1.1).
The finiteness of $\mathrm{cb}(\mathsf{p})$: let C be a canonical base of p . By 2) p is the unique nonforking extension of $p' = \mathsf{p}\restriction_C$. Whence p' has Morley degree 1 and contains an $L(C)$–formula with $\mathrm{MR}(\phi) = \mathrm{MR}(\mathsf{p})$ and $\mathrm{Md}(\phi) = 1$. Let $c \in \mathrm{dcl}^{\mathrm{eq}}(C)$ be the canonical base of ϕ. Since p is the unique nonforking extension of $\mathsf{p}\restriction_c$ we have $C \subset \mathrm{dcl}^{\mathrm{eq}}(c)$. This shows that c is also a canonical base of p. □

Lemma 4.13 *Let B be an extension of the model M and $p \in \mathrm{S}(M)$ definable over M. Then p has a unique extension to a type $q \in \mathrm{S}(B)$ which is still definable over M.*

Proof: If p is defined by the formulas $\mathrm{d}_p x\phi(x,\overline{y})$ the same formulas define an extension

$$q = \{\phi(x,\overline{b}) \mid \overline{b} \in B, \models \phi(x,\overline{b}))\}.$$

If r is a second M–definable extension of p then for all formulas $\phi(x,\overline{y})$

$$\mathrm{d}_q x\phi(x, M) = \mathrm{d}_p x\phi(x, M) = \mathrm{d}_r x\phi(x, M).$$

Since M is a model we have therefore $\mathrm{d}_q x\phi(x,\mathfrak{C}) = \mathrm{d}_r x\phi(x,\mathfrak{C})$ for all ϕ, which means $q = r$. □

Let p be a type over a model M and q an extension of p to some set B. We say that q is a *heir* of p if for every M–formula $\phi(x,\overline{b})$ in q there is a tuple $\overline{m}$ in M such that $\phi(x,\overline{m}) \in p$. We say that q is a *coheir* if every $\phi(x,\overline{b}) \in q$ is realized in M. It is easy to show without any assumption about T that every type p has a heir and a coheir on B.

Theorem 4.14 (Heirs and Coheirs) *Let p a type over a model M and q an extension of p to some set B. Then the following are equivalent:*

1. *q is a heir of p.*
2. *q is a coheir of p.*
3. *q is the nonforking extension of p.*

Proof: Let q be the nonforking extension of p. Then q is definable over M. $\phi(x,b) \in q$ implies that $\models \mathrm{d}_q x\phi(x,y)$ is realizable and then also realizable in M. Whence q is a heir. This shows 3→1.
If q' is another heir and $\neg\phi(x,b) \in q'$ then also $\neg\phi(x,b) \wedge \mathrm{d}_p\phi(x,b) \in q'$. But a formula $\neg\phi(x,m) \wedge \mathrm{d}_p\phi(x,m)$ cannot belong to p. Therefore $\phi(x,b) \in q'$ and we have shown that $q' = q$ and 1→3.
The equivalence of 3 and 1 with 2 follows from symmetry and the observation that $\mathrm{t}(a/Mb)$ is a heir of $\mathrm{t}(a/M)$ iff $\mathrm{t}(b/Ma)$ is a coheir of $\mathrm{t}(b/M)$. □

The notion of forking can be extended to stable theories (in a unique way). All properties of this section remain true with the following exceptions. The multiplicity is in general not finite but bounded by 2^{ω}. Also 4.5.1 is no longer true but q does not fork over a *countable* subset of B. The canonical base of a type can no longer be chosen to be a finite set.

We have given a quite complete list of the main properties of forking. Two further basic properties "Closedness" and "Open mapping" which are not needed for the purpose of this collection are postponed to the appendix.

5 Strongly minimal sets

Let T be an ω–stable theory and $\mathbb{D}$ a 0–definable strongly minimal class. Algebraic closure *restricted* to $\mathbb{D}$ maps every subset A of $\mathbb{D}$ to

$$\mathrm{acl}(A) \cap \mathbb{D}.$$

Theorem 5.1 *The algebraic closure operator turns* $\mathbb{D}$ *into a* pregeometry *in the sense of the following definition.*

Definition 5.2 *A set X together with an operator* $cl : \mathcal{P}(\mathcal{X}) \mapsto \mathcal{P}(\mathcal{X})$ *is a* pregeometry *if the following is true for all* $A \subset X$, $a, b \in X$

1. $A \subset cl(A)$
2. $cl(A) = \bigcup\{cl(A_0) \mid A_0 \text{ a finite subset of } A\}$
3. $cl(cl(A)) = cl(A)$
4. $a \in cl(A \cup \{b\}) \setminus cl(A) \Longrightarrow b \in cl(A \cup \{a\})$.

Proof of 5.1:
1, 2, 3 are easy to check. 4 can be proved directly from the definition, but it follows from forking symmetry since in a strongly minimal set $a \in \text{acl}(A \cup \{b\}) \setminus \text{acl}(A)$ is equivalent to $a \not\perp_A b$. □

A stationary type $p \in S(A)$ is *regular* if $(p(\mathfrak{C})), cl)$ is a pregeometry, where $p(\mathfrak{C}) = \{b \in \mathfrak{C} \mid b \text{ realizes } p\}$ and $cl(B) = \{b \in p(\mathfrak{C}) \mid b \not\perp_A B\}$. We have shown in theorem 5.1 that strongly minimal types are regular.

Example 1:

Let F be a division ring and V an F–vector space. If we define $cl(A)$ to be the span of A we obtain a pregeometry on V. If V is infinite this classical fact can be considered as a special case of 5.1: we consider V as an L_F–structure, where the language

$$L_F = \{0, +, -, \underline{r}\}_{r \in F}$$

contains unary function symbols $\underline{r}$ for multiplication by field elements. The theory T_F of infinite F–spaces is complete and strongly minimal and $\text{acl}(A)$ is the span of A.

Example 2:

An *affine space* is the second sort of a two–sorted structure (V, X), where V is a vector space which acts regularly on X. An affine subspace is either the empty set or of the form $x+W$, where W is a subspace of V. X becomes a pregeometry if one defines $cl(A)$ as the affine span of A, the smallest affine subspace which contains A. If X is infinite X is strongly minimal and $\text{acl}(A)$ is the affine span of A

Example 3:

If K is a field the relative algebraic closure (in the algebraist's sense) defines a pregeometry on K. Again, if K is algebraically closed, K is strongly minimal and the algebraic closure coincides with acl( ).

Let (X, cl) be a pregeometry. A set A is a *generating* set if $X = cl(A)$ and *independent* if $a \notin \mathrm{acl}(A \setminus \{a\})$ for all $a \in A$. A *basis* is a generating independent set.

Lemma 5.3 *Every pregeometry X has a basis. All bases have the same number of elements, the* dimension *of X,* $\dim(X)$.

Proof: If X is a vector space the lemma specializes to the basic theorem of linear algebra: every vector space has a basis and two bases have the same cardinality, the *dimension* of X. The standard proof of this theorem generalizes immediately to a proof of the lemma. □

Every subset Y of X gives rise to two new pregeometries:

1. Y itself is a pregeometry with the restricted closure operator $cl(A) \cap Y$.
2. The *localized* pregeometry X/Y on X with the relativized closure operator $cl_Y(A) = cl(A \cup Y)$.

If for example F is a subfield of the algebraically closed field K, then the dimension of F is the transcendence degree of F over its prime field and $\dim(K/F)$ is the transcendence degree of K over F.

One verifies easily the equation

$$\dim(X) = \dim(X/Y) + \dim(Y) \quad (*)$$

Let $\mathfrak{p}$ be the (strongly minimal) global type determined by $\mathbb{D}$. An element a realizes $\mathfrak{p}\restriction_A$ iff $a \in \mathbb{D}$ and $a \notin \mathrm{acl}(A)$. It follows that all elements of $\mathbb{D}$ which are not algebraic over A have the same type over A. From this follows by induction that elements $a_1 \ldots a_n$ of $\mathbb{D}$ which are (in the geometrical sense) independent over A have a unique type over A. We call the common (non–forking) global extension of these types $\mathfrak{p}^n$. This type has rank n by the following lemma.

Lemma 5.4 *If $a_1, \ldots, a_n \in \mathbb{D}$ are independent over A the type* $\mathrm{t}(a_1 \ldots a_n/A)$ *has rank n.*

Corollary 5.5 *For arbitrary $a_1, \ldots, a_n \in \mathbb{D}$*

$$\mathrm{MR}(a_1 \ldots a_n/A) = \dim(a_1 \ldots a_n/A).$$

Proof: The corollary follows from the lemma since by 2.9 $\mathrm{MR}(a_1 \ldots a_n/A) = \mathrm{MR}(a_1 \ldots a_k/A)$ if $a_{k+1}, \ldots, a_n$ are algebraic over $Aa_1 \ldots a_k$.

The lemma is clear for $n = 1$. Assume that the claim and (the corollary) is true for dimensions $\leq n-1$ and let $a_1, \ldots, a_n$ be independent over A. The induction hypothesis implies $\mathrm{MR}(a_1 \ldots a_n/Aa_1) = n-1$ and together with $a_1 \ldots a_n \not\downarrow_A a_1$ that $\mathrm{MR}(a_1 \ldots a_n/A) \geq n$. If $\mathrm{MR}(a_1 \ldots a_n/A) > n$ there would be two disjoint subclasses $\mathbb{E}$, $\mathbb{F}$ of $\mathbb{D}^n$ of rank n. Choose an extension B of A over which these classes are defined and n–tuples e and f from $\mathbb{E}$ and $\mathbb{F}$ which have rank n over

B. By induction the elements of e as well as the elements of f are independent over B. But then e and f have the same type over B, which is impossible. □

In the world of algebraically closed fields the corollary means that the dimension (in the sense of algebraic geometry) of a variety equals its Morley rank (see [Pil]).

Corollary 5.6 *In a strongly minimal set $\mathbb{D}$, Morley rank is definable. This means that for every formula $\phi(x_1, \ldots, x_n, y)$ which implies that the x_i's belong to $\mathbb{D}$ and for every k the class*

$$\{b \mid \mathrm{MR}(\phi(x_1, \ldots, x_n, b)) = k\}$$

is definable.

Proof: Since the rank of $\phi(x_1, \ldots, x_n, b)$ is bounded by n it is enough to show that for all k

$$\{b \mid \mathrm{MR}(\phi(x_1, \ldots, x_n, b)) \geq k\}$$

is definable. The rank of the formula $\phi(x_1, \ldots, x_n, b)$ is at least k iff it is realized by a sequence $a_1 \ldots a_n$ which contains a k–element subsequence (say $a_1 \ldots a_k$) which is independent over b (assuming that ϕ is an L–formula). Thus we have to show that

$$\mathrm{MR}(\exists x_{k+1} \ldots \exists x_n \phi(x_1, \ldots, x_n, b)) \geq k$$

is an elementary property of b. But this follows from the definability of $\mathfrak{p}^k$ and the fact that

$$\mathrm{MR}(\psi(x_1, \ldots, x_k, b)) \geq k \iff \psi(x_1, \ldots, x_k, b) \in \mathfrak{p}^k.$$

□

Definition 5.7 *A 0–definable class $\mathbb{H}$ is* almost strongly minimal *(with respect to the strongly minimal class $\mathbb{D}$) if it is contained in the algebraic closure of the 0–definable strongly minimal class $\mathbb{D}$.*

If $\mathfrak{C}$ is strongly minimal each sort of $\mathfrak{C}^{\mathrm{eq}}$ is an example of an almost strongly minimal class.

The rank of types of elements in $\mathbb{H}$ can be computed geometrically with the help of the following lemma.

Lemma 5.8 *For every $h \in \mathbb{H}$ there is an independent sequence $a_1, \ldots, a_m$, $b_1, \ldots, b_n$ in $\mathbb{D}$ such that h is independent from $a_1, \ldots, a_m$ and interalgebraic with $b_1 \ldots b_n$ over $a_1, \ldots, a_m$. We have then* $\mathrm{MR}(h/\emptyset) = n$.

Proof: Choose $B \subset \mathbb{D}$ minimal with $h \in \mathrm{acl}(B)$ and a basis $a_1, \ldots, a_m$ of B/h. □

Remark: For further applications we note that in the lemma any sequence $b_1 \ldots b_n$ which is independent over h can be used since all such sequences have the same type over h. Let I be an infinite independent subset of $\mathbb{D}$. If h is any element of H take a finite subset I_0 such that I is independent from h over I_0. Then $I \setminus I_0$ is independent over h and by the above h is interalgebraic with some finite tuple from $\mathbb{D}$ over $I \setminus I_0$. It follows that over I every element of $\mathbb{H}$ is interalgebraic with some finite tuple from $\mathbb{D}$. In particular, if we add enough parameters, every elemnt of $\mathbb{D}^{eq}$ is interalgebraic with a some tuple from the real sort $\mathbb{D}$ (in fact it is shown in [Pil] Lemma 1.6 that any strongly minimal set such that $\mathrm{acl}(\emptyset)$ is infinite admits weak elimination of imaginaries)

Now it is easy to generalize the dimension formula (*) to almost strongly minimal sets:

Corollary 5.9 (Lascar's equation) *For all* $g, h \in \mathbb{H}$

$$\mathrm{MR}(gh/F) = \mathrm{MR}(g/Fh) + \mathrm{MR}(h/F).$$

Proof: Find a set A independent from Fgh and tuples b and c from $\mathbb{D}$ which are interalgebraic with g and h over A. The ranks do not change if we add A to F and replace g, h by b, c. We then have to show that

$$\mathrm{MR}(bc/FA) = \mathrm{MR}(b/FAc) + \mathrm{MR}(c/FA).$$

This follows from (*). □

It is also easy to show that Morley rank is definable in almost strongly minimal sets. Actually Lascar's equation and definability of rank hold whenever each type is non–orthogonal (in the sense of the next section) to a strongly minimal type. For example if T is ω_1–categorical or if T is the theory of a group of finite Morley rank.

Definition 5.10 *A pregeometry* (X, cl) *is* modular *if for all cl-closed sets* A, B

$$\dim(A \cup B) + \dim(A \cap B) \quad = \quad \dim(A) + \dim(B) \qquad (**)$$

The pregeometry of a vector space is modular, affine spaces are not modular due to the existence of parallels.

The equation (**) is trivially true if A or B have infinite dimension. For finite dimensional subspaces the equation is equivalent to

$$\dim(B/A) \quad = \quad \dim(B/(A \cap B)). \qquad (***)$$

It is not hard to show that (***) still holds in modular geometries also for infinite dimensional A and B. This implies that all localizations of a modular pregeometry are modular.

Definition 5.11 *The strongly minimal class* $\mathbb{D}$ *is* locally modular *if for some sufficiently large subset* $D \subset \mathbb{D}$ *the localized geometry* $\mathbb{D}/D$ *is modular.*

Affine spaces are locally modular. It suffices to localize at a single point.

Let K be any algebraically closed subfield of the algebraically closed monster field $\mathfrak{C}$, of transcendence degree n. Choose three elements a, b, c independent over K. Take $\mathrm{acl}(K(a,b))$ for A and $\mathrm{acl}(K(c,ac+b))$ for B. Then $\dim(A) = \dim(B) = n+2$, $\dim(A \cup B) = \dim(\mathrm{acl}(K(a,b,c))) = n+3$ and $\dim(A \cap B) = \dim(K) = n$. This shows that algebraically closed fields are not locally modular. The next theorem gives an explanation: $ax + b \doteq y$ is a two parameter family of plane curves, which cannot exist in locally modular strongly minimal classes.

For simplicity we assume now that $\mathfrak{C}$ itself is strongly minimal. Corollary 3.7 explains how the following theory can be applied to arbitrary strongly minimal classes.

Theorem 5.12 *$\mathfrak{C}$ is locally modular iff every infinite family of plane curves is 1–dimensional. This means that for any A the canonical bases of all strongly minimal types $p(x,y) \in \mathrm{S}_2(A)$ have rank ≤ 1 over the empty set.*

Proof: Both parts of the equivalence remain the same if one adds a set D of parameters to the language. If for example the canonical base c of some $p \in \mathrm{S}_2(A)$ has rank $n \geq 2$ we may assume that A is independent from D: c is also a canonical base of the nonforking extension of p to AD and the rank of c over D is n.

Assume first that $\mathbb{C}$ is locally modular and even modular after adding parameters. If $p \in \mathrm{S}_2(A)$ is strongly minimal choose a realization $\bar{b}$. Let C be the intersection $\mathrm{acl}(\bar{b}) \cap \mathrm{acl}(A)$. Then by (**) $\mathrm{MR}(\bar{b}/C) = \mathrm{MR}(\bar{b}/A)$. This implies that p does not fork over C and therefore $\mathrm{cb}(p) \subset \mathrm{acl}^{\mathrm{eq}}(C)$. Whence it is enough to show that $\mathrm{MR}(C) \leq 1$. But this follows from Lascar's equation $\mathrm{MR}(\bar{b}/C) + \mathrm{MR}(C) = \mathrm{MR}(\bar{b}C) = \mathrm{MR}(\bar{b}) \leq 2$ and the fact $\mathrm{MR}(\bar{b}/C) = \mathrm{MR}(\bar{b}/A) = 1$.

For the converse assume that there is no rank≥ 2 family of plane curves. After adding parameters we can assume that all elements of $\mathfrak{C}^{\mathrm{eq}}$ are interalgebraic with tuples from C (see the remark after 5.8). We will prove that $\mathfrak{C}$ is modular. Claim: The pregeometry $\mathfrak{C}$ satisfies the modular law (**) if $\dim(B/(A \cap B)) = 2$. Proof: Let B be generated over $A \cap B$ by the pair $\bar{b}$. If (**) is false we have $\mathrm{MR}(\bar{b}/A) = 1$. Choose c of rank 1 which is interalgebraic with the canonical base of $p = \mathrm{t}(\bar{b}/A)$. Then $c \in A$ and since p forks over $\emptyset$ we have $\bar{b} \not\perp c$ and therefore $c \in B$. But this contradicts $\mathrm{MR}(\bar{b}/(A \cap B)) = 2$.

Finally we show that a pregeometry (X, cl) which has the property of the claim is modular. Let A, B be a counterexample to (**) with minimal dimension $n = \dim(B/(A \cap B))$. Let $b_1 \ldots b_n$ be a basis of B over $A \cap B$. Then $\mathrm{acl}(Ab_1 \ldots b_{n-2})$ and B are a counterexample to the hypothesis. □

A pregeometry is modular iff any two *cl*–closed sets A and B are (geometrically) independent over $A \cap B$. Formulated for $\mathfrak{C}^{\mathrm{eq}}$ this means that T is *one–based*.

Definition 5.13 *An ω–stable (or stable) theory is* one–based *if one of the following two equivalent conditions are satisfied in $\mathfrak{C}^{\mathrm{eq}}$ (!).*

1. *Any two* acl*–closed sets A and B are (forking–) independent over $A \cap B$.*
2. *The canonical base of any stationary type* $\mathrm{t}(a/B)$ *is algebraic over a.*

Note that by the proof of 3.2 the canonical base of a stationary type p is definable from a (in fact an independent) set of realizations of p.

That the two conditions are equivalent is clear from the following observations: 1. Condition 1 can be formulated for finite tuples instead of arbitrary sets. 2. In condition 2 one can assume that B is acl–closed. 3. For acl–closed A, B: the canonical base of $t(a/B)$ is contained in A iff $a \perp_{A \cap B} B$.

Theorem 5.14 *Let $\mathfrak{C}$ be strongly minimal. Then $\mathfrak{C}$ is locally modular iff it is one–based.*

Proof: If every element of $\mathfrak{C}^{eq}$ is interalgebraic with a tuple of elements of $\mathfrak{C}$ (which, by the remark after 5.8, we can achieve by adding parameters) the equivalence is obvious. Thus we have to show that the strength of one–basedness does not change if we add a parameter set D to the language.

If $\mathfrak{C}$ is one–based $\mathrm{acl}(AD)$ is of course independent from $\mathrm{acl}(BD)$ over $\mathrm{acl}(AD) \cap \mathrm{acl}(BD)$. Whence $\mathfrak{C}_D = (\mathfrak{C}, d)_{d \in D}$ is 1-based. For the converse assume that $\mathfrak{C}_D$ is one–based and look at the type $\mathrm{st}(a/B)$. We can assume that aB is independent from D, which implies that $\mathrm{st}(a/BD)$ is a nonforking extension of $\mathrm{st}(a/B)$. Let c be the canonical base of $t(a/B)$. Then c is the canonical base of $\mathrm{st}(a/BD)$ in the sense of $\mathfrak{C}_D$ and whence by our assumption is contained in $\mathrm{acl}(Da)$. Together with $c \perp_a D$ we get $c \in \mathrm{acl}(a)$. □

We have seen two classes of examples of strongly minimal sets: vector spaces and affine spaces which are locally modular, algebraically closed fields which are not. Another class is that of *trivial* strongly minimal sets: $\mathbb{D}$ is trivial if for any $A \subset \mathbb{D}$, $\mathrm{acl}(A) = \bigcup_{a \in A} \mathrm{acl}(\{a\})$. Trivial pregeometries are clearly modular. The typical examples of trivial strongly minimal sets are the infinite set with no structure (except equality) and $(\mathbb{Z}, succ)$, the integers with the successor function.

Note that a strongly minimal group cannot be trivial: if a, b are two non algebraic independent elements, then $c = a.b \in \mathrm{acl}(a, b) \setminus (\mathrm{acl}(\{a\}) \cup \mathrm{acl}(\{b\}))$.

For many years these were the only kind of examples of pregeometries known to arise from strongly minimal sets. This led naturally to the conjecture that there might actually not exist any radically different ones. This intuition turned out to be correct in the locally modular case: any non trivial locally modular strongly minimal set arises from a vector space over a division ring.

In the case of non locally modular, the following conjecture (known as Zilber's conjecture) remained open for a long time: let $\mathbb{D}$ be strongly minimal non locally modular; then $\mathbb{D}$ interprets an algebraically closed field (i.e. an algebraically closed field is definable in $\mathbb{D}^{eq}$). This was disproved by Hrushovski who constructed a non locally modular strongly minimal set which does not even interpret a group ([Hr 93]).

But as will be apparent later in this volume [Bous], the dichotomy locally modular/non locally modular is one of the main ingredients in Hrushovski's proof of the Mordell-Lang's conjecture: his proof uses in an essential way the fact that a strong version of Zilber's conjecture does indeed hold for a rich class of strongly minimal structures, the Zariski geometries. This is explained in this volume in [Mar].

For a fairly comprehensive survey of recent results about strongly minimal sets and geometries, see [Mar 97]. For a detailed exposition of the results on local modularity and one–based stable structures and more generally of geometric stability theory, see [Pi2 96].

6 Orthogonality

We still assume that T is an ω–stable theory and that $\mathfrak{C}$ is its monster model.

Definition 6.1 *Two definable classes* $\mathbb{D}$ *and* $\mathbb{E}$ *are* orthogonal[3] if any two elements $d \in \mathbb{D}$ and $e \in \mathbb{E}$ are independent over any set of parameters over which $\mathbb{D}$ and $\mathbb{E}$ can be defined. Notation: $\mathbb{D} \perp \mathbb{E}$.

The laws of forking imply that

$$d \underset{A}{\downarrow} e \iff d \underset{B}{\downarrow} e$$

if de and $B \supset A$ are independent over A. This rule implies that orthogonality fails for every extension B of A if it fails for the parameter set A.

The following characterization avoids "forking":

Lemma 6.2 $\mathbb{D}$ *and* $\mathbb{E}$ *are orthogonal iff for all* $d \in \mathbb{D}$ *and* $e \in \mathbb{E}$ *and all* acl^{eq}*–closed sets* A *over which* $\mathbb{D}$ *and* $\mathbb{E}$ *are defined*

$$\mathrm{t}(d/A) \cup \mathrm{t}(e/A) \vdash \mathrm{t}(de/A).$$

If $p(x)$, $q(y)$ and $r(x,y)$ are types over A the condition $p \cup q \vdash r$ means that all pairs $(d,e) \in p(\mathfrak{C}) \times q(\mathfrak{C})$ realize r.

Proof: The condition means that $\mathrm{t}(d/A)$ has only one extension (namely $\mathrm{t}(d/Ae)$) to a type over Ae, which must therefore be a (the) non–forking extension. This implies that all e, d are independent over all acl^{eq}–closed sets, which suffices for $\mathbb{D} \perp \mathbb{E}$. Conversely orthogonality implies that $\mathrm{t}(d/A)$ has only nonforking extensions to Ae, which are unique if A is acl^{eq}–closed. □

Lemma 6.3 *A strongly minimal class* $\mathbb{D}$ *is non–orthogonal to the definable class* $\mathbb{E}$ *iff* $\mathbb{D} \subset \mathrm{acl}(A \cup \mathbb{E})$ *for some (finite) parameter set* A.

[3] We use here only orthogonality of formulas and not of complete types. The reader should beware that this definition applied to types corresponds to the notion usually called "hereditarily orthogonal". The definition of orthogonality for types is the following: let $p, q \in S(A)$ be two complete stationary types; we say that p and q are orthogonal ($p \perp q$) if for any $B \supseteq A$, for any a realizing the (unique) non forking extension of p to B, for any b realizing the (unique) non forking extension of q to B, a and b are independent over B.

Proof: $\Leftarrow$: If $\mathbb{D} \subset \mathrm{acl}(A \cup \mathbb{E})$ choose $d \in \mathbb{D} \setminus \mathrm{acl}(A)$ and a minimal subset $\{e_1, \ldots, e_n\}$ of $\mathbb{E}$ such that $d \in \mathrm{acl}(A \cup \{e_1, \ldots, e_n\})$. Then $d \not\downarrow_{A \cup \{e_1, \ldots, e_{n-1}\}} e_n$.
$\Rightarrow$: $d \not\downarrow_A e$ for some A (over which $\mathbb{D}$ and $\mathbb{E}$ are defined) amounts to $d \in (\mathrm{acl}(Ae) \setminus \mathrm{acl}(A))$. Since all $d' \in \mathbb{D} \setminus \mathrm{acl}(A)$ have the same type over A we have $\mathbb{D} \setminus \mathrm{acl}(A) \subset \mathrm{acl}(A \cup \mathbb{E})$. □

Lemma 6.4 *Non-orthogonality is an equivalence relation for strongly minimal classes.*

Proof: Let the strongly minimal $\mathbb{D}$ be non-orthogonal to $\mathbb{E}_1$ and $\mathbb{E}_2$. If A is large enough we find $d_i \in \mathbb{D}$ and $e_i \in \mathbb{E}_i$ such that $d_i \in (\mathrm{acl}(Ae_i) \setminus \mathrm{acl}(A))$ $(i = 1, 2)$. Since d_1 and d_2 have the same type over A we can assume that $d_1 = d_2$ whence $d_1 \not\downarrow_A d_2$, which implies $e_1 \not\downarrow_A e_2$. This shows that $\mathbb{E}_1 \not\perp \mathbb{E}_2$. □

Non-orthogonality between an almost strongly minimal class $\mathbb{G}$ and definable class $\mathbb{E}$ becomes transparent if $\mathbb{G}$ carries the structure of an abelian group. In the proof of the next theorem we will use the following results from Lascar's paper on groups in this volume ([Las]): Call an ω-stable group *connected* if it has no proper definable subgroup of finite index.

- (2.6) Every ω-stable group $\mathbb{G}$ has a unique minimal definable subgroup $\mathbb{G}^0$ of finite index, its *connected* component.
- (4.3) G is a connected group if and only if it has only one generic type if and only if it has Morley degree 1.

Theorem 6.5 *(in T^{eq}) An infinite almost strongly minimal abelian group $\mathbb{G}$ is non-orthogonal to the definable class $\mathbb{E}$ iff there is a definable group $\mathbb{H} \subset \mathrm{dcl}^{\mathrm{eq}}\mathbb{E}$ and a definable surjective homomorphism $h : \mathbb{G} \to \mathbb{H}$ with finite kernel.*

Proof: If such $\mathbb{H}$ and h exist, then $\mathbb{G}$ is non-orthogonal to $\mathbb{H}$ and the result follows from

$$\mathbb{G} \perp \mathbb{E} \Longrightarrow \mathbb{D} \perp \mathbb{H}, \quad \text{if } \mathbb{H} \subset \mathrm{acl}^{\mathrm{eq}}(\mathbb{E}).$$

We first prove the other direction under the assumption that $\mathbb{G}$ has Morley degree 1. Let $\mathbb{G}$ be in the algebraic closure of the strongly minimal set $\mathbb{D}$ and non-orthogonal to $\mathbb{E}$. Then $\mathbb{D}$ is non-orthogonal to $\mathbb{E}$ too. Lemma 6.3 gives a finite set A of parameters such that $\mathbb{D}$ and therefore $\mathbb{G}$ are contained in $\mathrm{acl}(A \cup \mathbb{E})$. Choose A large enough, so that $\mathbb{G}$ and $\mathbb{E}$ are definable over A. We need the following lemma:

Lemma 6.6 *Let $\mathbb{E}$ be a 0-definable class and $g \in \mathrm{acl}(\mathbb{E})$. Then there is an element $f \in \mathrm{dcl}^{\mathrm{eq}}(g) \cap \mathrm{dcl}^{\mathrm{eq}}(\mathbb{E})$ such that $g \in \mathrm{acl}^{\mathrm{eq}}(f)$.*

Proof: Let g be contained in the finite class $\phi(\mathfrak{C}, \overline{e})$ for an n-tuple of elements of $\mathbb{E}$. We can find ϕ in such a way that $\phi(\mathfrak{C}, \overline{e'})$ has at most k elements for all $\overline{e'}$. Now take for f a canonical base of $\phi(g, \mathbb{E}^n)$. $f \in \mathrm{dcl}^{\mathrm{eq}}(g)$ is clear, $f \in \mathrm{dcl}^{\mathrm{eq}}(\mathbb{E})$ follows from 3.7. With the intention to prove that g has at most k conjugates over f assume that $g_0, \ldots, g_k$ are elements with the same type as g over f. Then all sets $\phi(g_i, \mathbb{E}^n)$ are equal to $\phi(g, \mathbb{E}^n)$. Since now all g_i lie in $\phi(\mathfrak{C}, \overline{e})$ not all of them can be different. This shows $g \in \operatorname{acl}(f)$. □

Now back to the proof of 6.5. We add A to the language and assume that A is empty. The lemma together with a compactness argument shows that there is a 0–definable function $h'' : \mathbb{G} \to \operatorname{dcl}^{\mathrm{eq}}(\mathbb{E})$ with finite fibers. h'' is our first approximation of h.

Let K be the definable subgroup (!)

$$\{a \in \mathbb{G} \mid h''(a + g') = h''(g') \quad \text{for one (=all) } g' \in \mathbb{G} \text{ which is generic over } a\}.$$

(Note that all generics have the same type over a since $\mathbb{G}$ has Morley degree 1.) K is finite since h'' has finite fibers. As an almost strongly minimal set $\mathbb{G}$ has finite rank r. Choose an independent sequence $g_0, \ldots, g_{2r}$ of generic elements and define as a second approximation $h' : \mathbb{G} \to \operatorname{dcl}^{\mathrm{eq}}(\mathbb{E})$ by

$$h'(a) = (h''(a + g_0), \ldots, h''(a + g_{2r})).$$

Claim: For all $b, c \in \mathbb{G}$

$$h'(b) = h'(c) \Longrightarrow b - c \in K.$$

Proof: If b and c are different mod K take an i such that g_i is generic over b, c. (This must exist since the pair bc has rank $\leq 2r$.) But then $c + g_i$ is generic over bc and whence $h''(b + g_i) = h''((b - c) + (c + g_i)) \neq h''(c + g_i)$. This show that $h'(b) \neq h'(c)$.

Define $h(b)$ as the set $\{h'(a + b) \mid a \in K\}$ considered as an element of $\operatorname{dcl}^{\mathrm{eq}}(\mathbb{E})$. Then clearly $h(b) = h(c) \iff b - c \in K$ and h transports the group structure from $\mathbb{G}/K$ to $\mathbb{H} = h[\mathbb{G}]$.

We still have to consider the case where $\mathbb{G}$ is not connected. Assume that $\mathbb{G}$ is non–orthogonal to $\mathbb{E}$. If the set A intersects every coset of $\mathbb{G}^0$, $\mathbb{G}$ is contained in $\operatorname{acl}(\mathbb{G}^0 \cup A)$. This shows that $\mathbb{G}^0$ is non–orthogonal to $\mathbb{E}$. We have already shown that there is a finite normal subgroup K of $\mathbb{G}^0$ such that $\mathbb{G}^0/K$ is definably isomorphic to a definable group $\mathbb{H}_0$ which lives in $\operatorname{dcl}^{\mathrm{eq}}(\mathbb{E})$. Since the centralizer of K in $\mathbb{G}^0$ has finite index and $\mathbb{G}^0$ is connected K is contained in the center of $\mathbb{G}^0$. On the other hand K has only finitely many conjugates in $\mathbb{G}$. It follows that the product of these conjugates is a finite normal subgroup $\overline{K}$ of $\mathbb{G}$. $\mathbb{G}^0/\overline{K}$ is definably isomorphic to a quotient $\overline{\mathbb{H}}_0$ of $\mathbb{H}_0$ by a finite normal subgroup. $\overline{\mathbb{H}}_0$ still lives in $\operatorname{dcl}^{\mathrm{eq}}(\mathbb{E})$. If we extend the isomorphism to an isomorphism h between $\mathbb{G}/\overline{K}$ and an extension $\mathbb{H}$ of $\overline{\mathbb{H}}_0$, at first sight the group $\mathbb{H}$ does not seem to be definable. But after a choice of representatives for the cosets of $\mathbb{G}^0$ in $\mathbb{G}$ the elements of $\mathbb{G}/\overline{K}$ can be coded by elements of $\mathbb{G}^0/\overline{K}$. h transports this coding to $\mathbb{H}$, which shows that $\mathbb{H}$ can be found to live in $\operatorname{dcl}^{\mathrm{eq}}(\mathbb{E})$ and h to be a definable map. □

A final remark: By 3.7 $\mathbb{H}$ is definable with parameters from $\mathbb{E}$ (and the parameters used for $\mathbb{E}$). Furthermore, if we have elimination of imaginaries, we find such a definable $\mathbb{H}$ included in $\mathbb{E}$.

7 Appendix

We now complete the list of the main properties of forking, which we started section 4.

S(A) is the Stone space of the boolean algebra of A–definable classes. It is a compact space with basic (cl)open sets $\{p \mid \phi \in p\}$ for all $L(A)$–formulas $\phi(x)$. The next theorem says that the set of all types over B which do not fork over A is a closed subset of S(B).

Theorem 7.1 (Closedness) *If the type $q \in S(B)$ forks over A it must contain a* forking *formula. That is a formula which belongs only to types over B which fork over A.*

Proof: Since the set of nonforking $q \in S(B)$ is the continuous image of the set of global types which do not fork over A it is enough to prove that this set is closed in S($\mathfrak{C}$). This allows us in turn to assume that A is algebraically closed in $\mathfrak{C}^{eq}$. If q forks over A there is a finite tuple b and an element a such that $a \not\perp_A b$ and $t(a/Ab) \subset q$. Let r be the global nonforking extension of $t(b/A)$. Since $t(b/Aa) \not\subset r$ there is a formula $\phi(x,y)$ over A such that $\models \phi(a,b)$ and $\not\models d_r y\phi(a,y)$. The formula

$$\phi(x,b) \wedge \neg d_r y\phi(x,y)$$

forks and belongs to q. □

Theorem 7.2 (Open mapping) *Let* NF(B/A) *be the set of types over B which do not fork over A. Then the restriction map* NF(B/A) $\to$ S(A) *is open.*

Proof: Again it is enough to show that $\pi : \mathrm{NF}(\mathfrak{C}/A) \to S(A)$ is open. If O is an (relatively) open subset of NF($\mathfrak{C}/A$) the inverse image I of $\pi(O)$ is the union of all conjugates of O and therefore open. But then $S(A) \setminus \pi(O) = \pi(\mathrm{NF}(\mathfrak{C}/A) \setminus I)$ is closed and $\pi(O)$ is open. □

References

[Bous] E. Bouscaren, *Proof of the Mordell-Lang conjecture for function fields*, this volume.

[De] F. Delon, *Separably closed fields*, this volume.

[Hr 93] E.Hrushovski, *A new strongly minimal set*, Ann. Pure Appl. Logic 62 (1993), 147-166.

[Las 87] D. Lascar, *Stability in Model Theory*, Longman Scientific and Technical, New York, 1987.

[Las] D. Lascar, *ω-stable groups*, this volume.

[Mar] D. Marker, *Zariski geometries*, this volume.

[Mar 97] D. Marker, *Strongly minimal sets and geometry*, to appear in Logic Colloquium '95 (Haifa).

[Pi2 96] A. Pillay, *Geometrical Stability Theory*, Oxford University Press, 1996.

[Pi1] A. Pillay, *Model theory of algebraically closed fields*, this volume.

[Wo] C. Wood, *Differentially closed fields*, this volume.

Omega-stable groups

Daniel Lascar

This is a self-contained presentation of the theory of ω-stable groups. All proofs are given.

Many of the results are valid with the weaker assumption of stability instead of ω-stability, but in order to remain self-contained we present here only the ω-stable case and make full use of this strong assumption in order to keep the proofs as simple as possible.

Properties of ω-stable groups play a fundamental role in applications of model theory to algebraic geometry. This is not very surprising as any algebraic group (over an algebraically closed field) is an ω-stable group (see the chapter on algebraically closed fields [Pi1]).

In fact, a long standing conjecture in the subject is that any simple ω-stable group is an algebraic group over an algebraically closed field (see [BoNe] or [Po 87]).

1 Prerequisites and notation

We list here some notation and prerequisites, most of which already appear in the introductory article of Ziegler, in this volume ([Zie]).

1.1 When we say definable without further precision, we mean definable with parameters.

1.2 We will use the Morley rank and degree of a set and of a type. We will denote the Morley rank and the degree of the type of a tuple of elements $\bar{a}$ over a set A respectively by $MR(\bar{a}/A)$, $d(\bar{a}/A)$ (see [Zie], section 2).

1.3 Every complete type over a model is definable and has a canonical extension called its heir (see [Zie], Corollary 3.5 and section 4).

1.4 If X is a set definable with parameters in A, then

$$MR(X) = \max\{MR(\bar{a}/A)\ ;\ \bar{a} \in X\}.$$

(see [Zie], section 2).

1.5 If $(X_i\ ;\ i \in I)$ is a finite partition of X, then $MR(X) = \max(MR(X_i)\ ;\ i \in I)$, and if $MR(X_i)$ is constant, $d(X) = \sum_{i\in I} d(X_i)$ (see [Zie], Lemma 2.5).

1.6 If f is a definable injective map and X is a definable set, then $MR(X) = MR(f[X])$ (see [Zie], 2.10).

1.7 If M is a model and X is a definable subset of M^n, then $X[\bar{v}]$ (with the right number of free variables) is the formula defining it. If $F[\bar{v}] = F$ is a formula, then $F[M]$ is the subset of M (or of M^n) defined by $F[\bar{v}]$.

1.8 We will use freely imaginary elements, the canonical base of a set or of a type (see [Zie], sections 1 and 4).

1.9 Symmetry lemma (Theorem 4.7 of [Zie]): If $\bar{a}$ and $\bar{b}$ are tuples of elements and if A is a set, then $MR(\bar{a}/A \cup \{\bar{b}\}) = MR(\bar{a}/A)$ if and only if $MR(\bar{b}/A \cup \{\bar{a}\}) = MR(\bar{b}/A)$.

We will assume that G is an ω-stable group, that is a group with possibly extra structure such that the theory of G with the extra structure is ω-stable. Everything that is said here extends without any difficulties to the situation where G is a group which is definable or even interpretable in an ω-stable structure. The only precaution one should take in this case is to consider the parameters necessary to define or to interpret the group as individual constants and part of the language.

2 Chain conditions

Our aim in this section is to prove that, in an ω-stable group, theré is no infinite decreasing chain of definable subgroups, and to deduce some consequences of this fact.

Lemma 2.1 *Assume that H is a definable subgroup of G. Then for any $a \in G$,*

$$MR(H) = MR(aH) = MR(Ha)$$

Proof: By 1.6, because there is a definable bijection from H to aH. □

Corollary 2.2 *Assume that $H \subset H'$ are definable subgroups of G.*

1. *If $[H' : H]$ is finite, $MR(H) = MR(H')=$ and $d(H') = d(H) \cdot [H'/H]$.*
2. *If $[H' : H]$ is infinite, then $MR(H') > MR(H)$.*

Corollary 2.3 *There is no infinite decreasing sequence of definable subgroups.*

Corollary 2.4 *The intersection of any class of definable subgroups is equal to the intersection of a finite number of them, and consequently, is definable.*

As a matter of fact, this corollary has two kinds of applications, that we are going to illustrate now.

Example : Let A be any subset of G (not necessarily definable). Then the centralizer of A

$$Z_A = \{g \in G \; ; \text{ for all } a \in A,\, ga = ag\} = \bigcap \{Z_a \; ; \; a \in A\}$$

is the centralizer of a finite subset of A. So it is definable.

Example : Assume that $F[v_0, v_1]$ and $H[v_0]$ are formulas, and that for all $a \in G$ satisfying H, $F[a, G]$ is a subgroup of G. The intersection of all subgroups $F[a, G]$ with $a \in H[G]$ is of course definable by the formula

$$K[v_0] = \forall v_1 (H[v_1] \Longrightarrow F[v_1, v_0]).$$

But this group is also the intersection of a finite number of subgroups $F[a, G]$, with $a \in H[G]$. Moreover, $K[v_0]$ defines the same intersection in any $G' \succeq G$.

For example, let A be a subset of G definable by the formula $A[v_0]$. Then, Z_A, the centralizer of A, is definable by the formula

$$Z_A[v_0] = \forall v_1 (A[v_1] \Longrightarrow v_0 v_1 = v_1 v_0)$$

and for any $G' \succeq G$, the centralizer of the set $A[G']$, the subset of G' defined by the formula $A[v_0]$ is equal to $Z_A[G']$. But, in fact, corollary 2.3 implies that there is a finite set $A_0 = \{a_1, a_2, \ldots, a_n\}$ of G, $A_0 \subset A$ such that $Z_A = Z_{A_0}$. Thus

$$G \models \forall v_0 (\forall v_1 (A[v_1] \Longrightarrow v_0 v_1 = v_1 v_0) \Longleftrightarrow \bigwedge_{i=1}^{i=n} a_i v_0 = v_0 a_i)$$

Now, if G' is an elementary extension of G (or an elementary restriction containing all the relevant parameters), then

$$G' \models \forall v_0 (\forall v_1 (A[v_1] \Longrightarrow v_0 v_1 = v_1 v_0) \Longleftrightarrow \bigwedge_{i=1}^{i=n} a_i v_0 = v_0 a_i)$$

and this prove that the centralizer of $A[G']$ is also equal to the centralizer of A_0 in G'.

Consider now the class of definable subgroups of G of finite index. The intersection of all these subgroups is the intersection of a finite number of them, so it is itself a definable subgroup of G of finite index. In other words, there is a unique minimal definable subgroup of finite index.

Definition 2.5 *Let H be a definable subgroup of G. We say that H is* connected *if it has no proper definable subgroup of finite index.*

Remark: It is immediate to check that if G is connected and H is a normal definable subgroup of G, then G/H is connected.

Definition 2.6 *The* connected component *of G is the intersection of all definable subgroups of G of finite index. It is denoted G^0.*

So, we have proved that G^0 is a definable subgroup of G of finite index. It is connected. We note that $MR(G) = MR(G^0)$. If $d(G) = 1$, then G is connected (use Corollary 2.2). We will see the converse in the next section.

Proposition 2.7 *G^0 is definable by a formula without parameters, that we shall denote $G^0[v_0]$. If G' is an elementary extension of G, then $G^0[G']$ is the connected component of G'.*

Proof: Let $F[x, \bar{a}]$ be the formula ($F[x, \bar{y}]$ a formula without parameters and $\bar{a}$ a sequence of elements in G) such that $G^0 = \{g \in G\ ;\ G \models F[g, \bar{a}]\}$. Let $k = [G : G^0]$. It is routine to write a formula $H[\bar{y}]$ such that: for all $\bar{b}$ in G, $G \models H[\bar{b}]$ if and only if the set $F[G, \bar{b}]$ is a subgroup of G of index k, in other words is equal to G^0. Then G^0 is defined by the formula $G^0[x] = \exists \bar{y}(H[\bar{y}] \wedge F[x, \bar{y}])$.

For every formula $H[x]$ and for every integer n, there is a formula expressing that $H[G]$ is a subgroup of index n. To express that G^0 is connected, it suffices to write, for every formula $H[x, \bar{y}]$ (without parameters) and integer n: $\neg\exists \bar{y}(H[x, y] \wedge G^0[x]$ *defines a subgroup of finite index* n).

If G' is an elementary extension of G, these statements are also true in G', and $G^0[G']$ is the connected component of G'. □

The proof yields also:

Proposition 2.8 *Assume that G is connected and that G' is elementarily equivalent to G. Then G' is connected.*

3 Stabilizers

The group G acts naturally on $S_1(G)$, the space of one-types over G: let $p \in S_1(G)$ and $g \in G$. Then, by definition

$$g \cdot p = \{F[x]\ ;\ F \text{ is a formula with parameters in } G \text{ and } F(g \cdot x) \in p\}.$$

The type $g \cdot p$ could be defined by: if a is an element in an elementary extension of G realizing p, then $g \cdot p = t(ga/G)$. This remark shows that $MR(p) = MR(g \cdot p)$ (by 1.6). We leave it to the reader to check that it is a group action: $(gg') \cdot p = g \cdot (g' \cdot p)$= and $1 \cdot p = p$.

Definition 3.1 *Let $p \in S_1(G)$. The stabilizer of p, denoted $stab_p$, is the subgroup*

$$\{g \in G\ ;\ g \cdot p = p\}.$$

Proposition 3.2 *1. For any $p \in S_1(G)$, $stab_p$ is a definable subgroup of G.*

2. If G' is an elementary extension of G and p' is the heir of p over G', then $stab_{p'}$ is the subgroup of G' defined by $stab_p[x]$, that is,

$$stab_{p'} = \{g \in G'\ ;\ G' \models stab_p[g]\}.$$

Proof: 1. Let $F[x]$ be a formula with parameters in G (x a single variable) such that $F[x] \in p$ and p is the unique type of rank $MR(p)$ containing the formula $F[x]$. We first note that, for all $g \in G$, $g \in stab_p$ if and only if $F[gx] \in p$. One direction follows immediatly from the definition, and, assuming that $F[gx] \in p$, we see that $F[x] \in g \cdot p$. Since $MR(g \cdot p) = MR(p)$, $p = g \cdot p$ by choice of $F[x]$.

We now use the fact that p is definable (see 1.3): there exists a formula $H[y]$ such that, for any $g \in G$, $F[g \cdot x] \in p$ if and only if $G \models H[g]$. Thus the formula $H[y]$ defines the set $stab_p$ in G.

2. We know (1.3) that p' has the same rank as p and is the unique type in $S_1(G')$ of rank $MR(p)$ and containing $F[x]$. So, $stab_{p'} = \{g \in G' \; ; \; F(g \cdot x) \in p'\}$, as above. We also know (1.3) that the definition of p' is the same as the definition of p, so $F[gx] \in p'$ if and only if $G' \models H[g]$. □

Proposition 3.3 *For any* $p \in S_1(G)$, $MR(stab_p) \leq MR(p)$.

Proof: Let a be a realization of p and b be such that $\models stab_p[b]$, a and b independent over G (a and b in an elementary extension of G). Let G' be an elementary extension of G containing b and such that $t(a/G')$ is equal to p' the heir of p over G'. Since $b \in stab(p')$, $t(ba/G') = p'$. But $MR(ba/G \cup a) = MR(b/G \cup a)$ by 1.6, $MR(b/G \cup a) = MR(b/G)$ since a and b are independent over G; $MR(ba/G \cup a) \leq MR(ba/G) = MR(p)$. (because $t(ba/G) = p$).

So, for any $b \in stab_p$, $MR(b/G) \leq MR(p)$, and this proves, by 1.4 that $MR(stab_p) \leq MR(p)$. □

4 Generic types

Definition 4.1 *We say that* $p \in S_1(G)$ *is* generic *if* $MR(p) = MR(G)$.

As immediate consequences of this definition, generic types always exist, and there are exactly $d(G)$ different generic types; the heir of a generic type is generic; if $p \in S_1(G)$ is generic and $g \in G$, then $g \cdot p$ is also generic.

Theorem 4.2 *A type* $p \in S_1(G)$ *is generic if and only if* $stab_p$ *has finite index in* G.

Proof: If $stab_p$ has finite index in G, then $MR(stab_p) = MR(G)$ (by 1.5) and, by 3.3, $MR(p) \geq MR(G)$. Thus $MR(p) = MR(G)$ and p is generic.

Conversely, assume that p is generic. Since there are only a finite number of types in $S_1(G)$ of rank $MR(G)$, the orbit of p under the action of G:

$$\{g \cdot p \; ; \; g \in G\}$$

is finite. But for g, g' in G, $g \cdot p = g' \cdot p$, if and only if $g'^{-1}g \in stab_p$, and this implies that the index of $stab_p$ in G is equal to the cardinality of the orbit of p. □

Proposition 4.3 *G is a connected group if and only if it has only one generic type (if and only if its degree is 1).*

Proof: It is clear, as we already noted previously that if $d(G) = 1$, then G is connected. Conversely, assume that G is connected. Let p and p' be generic types. We are going to prove that $p = p'$. Since the stabilizers of p and p' are subgroups of G of finite index, it follows from the fact that G is connected that $stab_p = stab(p') = G$.

Let a and a' be independent realizations of p and p' respectively. Let G_1 be an elementary extension of G containing a' and such that $t(a/G_1) = p_1$ is the heir of p. Since $stab(p_1) = G_1$, we see that $t(a'a/G_1) = p_1$, and $t(a'a/G) = p$.

The above argument proves that, if b and b' are independant realizations of generic types, then $t(b'b/G) = t(b/G)$. This can be applied to a^{-1} and a'^{-1}, which are also independant realizations of generic types, as is easily seen. So $t(a^{-1}a'^{-1}/G) = t(a'^{-1}/G)$, and it follows that $t(a'a/G) = p' = p$. □

Proposition 4.4 *For any $p \in S_1(G)$, p is generic if and only if $stab_p = G^0$.*

Proof: As usual, one direction is obvious: if $stab_p = G^0$, then by Theorem 4.2, p is generic

Conversely, assume that p is generic. We have already seen that, in this case, $stab_p$ has finite index in G, so $G^0 \subseteq stab_p$. Let $a \in stab_p$, and let c be a realization of p. First, since G^0 has finite index in G, there is a finite number of elements of G, $c_1, c_2, \ldots, c_k$ such that:

$$G \models \forall x \bigvee_{i=1}^{i=k} G^0 \left[xc_i^{-1}\right]$$

The same formula being true in the monster model, we deduce that there is one of the c's, say it is c_1, such that $G^0 \left[cc_1^{-1}\right]$. But, since $a \in stab_p$, $t(ac/G) = t(c/G)$, so $G^0 \left[acc_1^{-1}\right]$. It follows that $a \in G^0$. □

Corollary 4.5 *The degree of G is equal to $\left[G : G^0\right]$; for any two generic types p and p' there exists $g \in G$ such that $g \cdot p = p'$.*

Proof: The first assertion follows from the fact that the degree of G^0 is one (because G^0 is connected). If p is generic, the cardinality of its orbit is exactly equal to $[G : stab_p] = \left[G : G^0\right]$, that is exactly the number of generic types. So, there is only one orbit of generic types. □

Proposition 4.6 *Let H be a connected definable subgroup of G and p the generic type of H. Then*

1. *Let G' be a $|G|^+$-saturated elementary extension of G. Then in G' any element of H is the product of two realizations of p.*

2. If X is a definable subset of H and $MR(X) = MR(H)$, then $H = XX$ (that is, every element of H is the product of two elements of X).

Proof: 1. Let $a \in H$ and c a realization of p. Then $MR(ac^{-1}/G) = MR(c/G)$ (because, over G, c is definable over ac^{-1} and *vice versa*), thus $t(ac^{-1}/G)$ is a generic of H, and it is equal to p. But $a = ac^{-1}c$.

2. Since the degree of H is 1, it is impossible that both X and $H - X$ have rank $MR(H)$. Thus $\neg X(x) \notin p$, and $X(x) \in p$. It follows that for any $a \in H$,

$$\mathcal{C} \models \exists v_0 \exists v_1 (a = v_0 v_1 \wedge X(v_0) \wedge X(v_1))$$

and the same formula holds in G. □

We conclude this section by a useful fact about groups which are definable or infinitely definable in an ω-stable theory. Let us first define these notions precisely.

We are given a complete theory T, in a language L. A group definable in T is a formula $\psi(v_1, \dots, v_n)$, together with two definable functions μ (2n-ary) and ι (n-ary) such that for any model M of T, the set $\psi(M)$ endowed with the restriction of these two functions μ and ι is a group G (μ being the multiplication of G and ι the inverse function). Now, an infinitely definable group is a family $\{\varphi_i(v_1, \dots, v_n); i \in I\}$ of first order formulas, together with two definable functions μ and ι such that for any $\mid I \mid^+$-saturated model M of T, the set $\cap_{i \in I} \varphi_i(M)$ endowed with the restriction of these two functions μ and ι is a group G. The main theorem is:

Lemma 4.7 *Assume that G is a group which is infinitely definable in an ω-stable theory. Then G is definable.*

Proof: To simplify notation, suppose that G is an infinitely definable subset of M, that is $G = \bigwedge_{i \in I} \varphi_i(v_0)$. We denote by $a \cdot b$ the product in G. So, there is a definable function from M^2 into M, denoted $\cdot$ such that, for any a, b in G, $a \cdot b \in G$, and a definable function from M to M denoted v_0^{-1} such that, for any $a \in G$, $a^{-1} \in G$. Moreover, we will assume, without loss of generality that the set $\{\varphi_i(v_0); i \in I\}$ is closed under logical consequences.

We see that $\{\varphi_i(v_0); i \in I\} \cup \{\varphi_i(v_1); i \in I\} \cup \{\varphi_i(v_2); i \in I\}$ implies

1. $v_0 \cdot 1 = 1 \cdot v_0 = v_0$

2. $v_0^{-1} \cdot v_0 = v_0 \cdot v_0^{-1} = 1$

3. $(v_0^{-1})^{-1} = v_0$

4. $(v_0 \cdot v_1) \cdot v_2 = v_0 \cdot (v_1 \cdot v_2)$

By compactness, there is an index $i \in I$, say $i = 0$ to simplify notation, such that $\varphi_0(v_0)$ implies 1), 2) and 3), and such that $\varphi_0(v_0) \wedge \varphi_0(v_1) \wedge \varphi_0(v_2)$ implies 4). We may moreover assume that the types in $\varphi_0(v_0)$ of maximal Morley rank are in G and, replacing $\varphi_0(v_0)$ by $\varphi_0(v_0) \wedge \varphi_0(v_0^{-1})$, that $\varphi_0(v_0)$ is equivalent to $\varphi_0(v_0^{-1})$.

Let A be a finite set containing the parameters of φ_0 and let

$$P = \{p \in S_1(A); \varphi_0(v_0) \in p \text{ and } RM(p) = RM(\varphi_0(v_0))\}$$

(so P is a finite subset of $S_1(A)$ and any realisation of any element of P belongs to G). Consider

$H = \{x \in M;$ for all $a \in M$ such that $t(a/A) \in P$ and a is independent from x over A, $M \models \varphi_0(x \cdot a)\}$.

We first see that $H \subseteq G$: let $x \in H$ and a a realisation of $p \in P$ which is independent from x over A. Then $RM(x \cdot a/A) \geq RM(x \cdot a/A \cup \{x\}) = RM(a/A \cup \{x\}) = RM(a/A) = RM(\varphi_0(v_0))$. Thus $x \cdot a \in G$, and $x = x \cdot (a \cdot a^{-1}) = (x \cdot a) \cdot a^{-1} \in G$.

Conversely, if $x \in G$ and a realises $p \in P$ and is independent from x over A, then $x \cdot a \in G$. So $M \models \varphi_0(x \cdot a)$ and $x \in H$.

So, it remains to prove that H is a definable set. But

$$H = \{b; \varphi_0(v_0) \subseteq_\alpha \varphi_0(v_0 \cdot b)\},$$

(see [Zie] Theorem 3.2) this is a definable set. □

5 The indecomposability theorem

In this section, we assume that G is an ω-stable group *of finite Morley rank.*

Definition 5.1 *Let X be a definable subset of G. We say that X is* indecomposable *if, for every definable subgroup H of G, the set $\{xH \; ; \; x \in X\}$ is either infinite or of cardinality 1.*

With the above notation, we will denote by X/H the set $\{xH \; ; \; x \in X\}$.

An example of an indecomposable subset is a connected definable subgroup of G.

Remarks:

1. Let X be a strongly minimal set. Then there exists an indecomposable set X_0 included in X such that $X - X_0$ is finite. Consider the family:

$$\mathcal{K} = \{K \; ; \; K \text{ is a subgroup of } G \text{ and } X/K = \text{ is finite}\}$$

and

$$K_0 = \bigcap_{K \in \mathcal{K}} K .$$

We know by corollary 2.3 that K_0 is the intersection of a finite family of $\mathcal{K}$, so X/K_0 is finite. Since X is strongly minimal, one and only one of the sets $xK_0 \cap X$ for $x \in X$ is infinite, and its complement in X is finite. Call X_0 this infinite set, then X_0 is indecomposable. Indeed, if K is a subgroup of G and

X/K is infinite, then X_0/K is also infinite, since $X - X_0$ is finite. On the other hand, if X/K is finite, then $K_0 \subseteq K$ and X_0/K has only one element.
2. Suppose that a subset X of G is normal (i.e., for any $g \in G$, $gXg^{-1} = X$). Then, to check that it is indecomposable, it suffices to prove that, for any *normal* subgroup H of G, the set X/H is either infinite or of cardinality 1. Indeed, let K be a definable subgroup of G and assume that the cardinality of X/K is finite, greater than 1. For all $g \in G$, let $K^g = \{gkg^{-1} ; k \in K\}$ be the g-conjugate of K. Since g-conjugation leaves X fixed and sends K to K^g, we see that $card(X/K^g) = card(X/K)$. By theorem 2.3, we know that there exists a finite subset A of G such that

$$K' = \bigcap_{g \in G} gKg^{-1} = \bigcap_{g \in A} gKg^{-1}$$

If $x \in X$, in order to determine its class modulo $\bigcap_{g \in A} gKg^{-1}$, it suffices to know its class modulo gKg^{-1} for all $g \in A$, and this proves that X/K' is finite, greater than 1.

It is the following theorem, due to Zilber and generalizing a crucial fact in the theory of algebraic groups, which makes indecomposable sets so interesting.

Theorem 5.2 The indecomposability theorem *Let $(X_i \; ; \; i \in I)$ be a set of indecomposable definable subsets of G, and assume that each of them contains the identity e of G. Then the group H generated by $\bigcup_{i \in I} X_i$ is definable and connected.*

Proof: Without loss of generality, we assume that, for any $i \in I$, there exists $j \in I$ such that $X_i^{-1} = X_j$ (we replace the set $(X_i \; ; \; i \in I)$ by the set $(X_i \; ; \; i \in I) \cup (X_i^{-1} \; ; \; i \in I)$). For each finite sequence $s = (i_1, i_2, \ldots i_k)$ of elements of I, consider the definable subset

$$X_s = \{x_1 x_2 \cdots x_k \; ; \; x_1 \in X_{i_1}, x_2 \in= X_{i_2}, \ldots, x_k \in X_{i_k}\}$$

Of course, X_s is included in H. Let $t = (j_1, j_2, \ldots j_k)$ be such a sequence, such that the rank of X_t is maximum (here we use the fact that the rank of G is finite). Let $X_t[x]$be the formula with parameters in G defining X_t, and let $p \in S_1(G)$ be a type such that $X_t[x] \in p$ and $MR(p) = MR(X_t[x]) = m$.

I claim that $H = stab_p$, and I first show that $H \subseteq stab_p$. For this, it suffices to prove that, for all $i \in I$, $X_i \subseteq stab_p$. Assume not. Since $e \in X_i$, $stab_p \in \{xstab_p \; ; \; x \in X_i\}$; since X_i is not included in $stab_p$, $stab_p$ is not the unique element of the set $\{xstab_p \; ; \; x \in X_i\}$; since X_i is indecomposable, we see that there is an infinite set of elements in X_i, $\{a_n \; ; \; n \in \omega\}$ such that, for any two distinct n, n' in ω, $a_n stab_p \neq a_{n'} stab_p$, so $a_n \cdot p \neq a_{n'} \cdot p$. Then, all the types $a_n \cdot p$ have rank m. They all contain the formula

$$F[x] = \exists v_0 \exists v_1 (X_i[v_0] \wedge X_t[v_1] \wedge x = v_0 v_1)$$

which is the formula defining the set $X_i X_t$. Thus this set has rank strictly bigger than m and this contradicts the maximality of m.

Now that we know that $H \subseteq stab_p$, we see that $X_t \subseteq stab_p$. But, by 3.3, $MR(stab_p) \leq MR(p) = MR(X_t)$, thus $MR(X_t) = MR(stab_p)$. This implies that $stab_p$ is connected: indeed, p is a generic type of the group $stab_p$, and it follows from proposition 4.4 that the stabilizer of p in $stab_p$, which is of course equal to $stab_p$, is the connected component of $stab_p$.

We may apply 4.6, and we get that $stab_p = X_t X_t$, and $stab_p \subseteq H$. □

Exactly as in the case of theorem 2.3, we can apply this theorem to a fixed infinite family $(X_i \; ; \; i \in I)$ of indecomposable sets. In this case, the subgroup generated by these formulas is in fact generated by a finite subfamily of $(X_i \; ; \; i \in I)$, and is definable. But we can also apply this theorem to a definable family of indecomposable sets, parametrized by a definable subset of G^n. Here is an example.

Example Assume G is connected. Then the derived group G', generated by the commutators $\{(aba^{-1}b^{-1}) \; ; \; a, b \in G\}$ is definable and connected.

For each $a \in G$, consider the set $X_a = \{(aba^{-1}b^{-1}) \; ; \; b \in G\}$. It is clear that G' is the subgroup generated by $\bigcup_{a \in G} X_a$. It is also clear that $e \in X_a$, for all $a \in G$. Let's prove that X_a is indecomposable.

Set $Y_a = a^{-1}X_a = \{ba^{-1}b^{-1} \; ; \; b \in G\}$. For any subgroup H of G, we have: $card(X_a/H) = card(Y_a/H)$ (because right multiplication by a^{-1} induces a bijection between these two sets), so it suffices to show that Y_a is indecomposable. By Remark 2 above, it suffices to prove that, if H is a normal subgroup of G, then $card(X_a/H)$ is either 1 or infinite. Now, we see that $ba^{-1}b^{-1}H = ca^{-1}c^{-1}H$ if and only if in the quotient group G/H, bc^{-1} belongs to the centralizer of aH. In other words, $card(X_a/H)$ is equal to the index of this centralizer. It cannot be finite greater than 1 because G/H is connected (remark 2).

Remark In fact, to get the definability of the derived group, it is not necessary to assume that G is connected. We leave this as an exercise.

6 One-based groups

Definition 6.1 *An ω-stable theory is said to be* one-based *if for any $n \in \omega$, any model M of T, any complete type $p \in S_n(M)$, and any realization $\bar{a}$ of p, $Cb(p)$ is algebraic over $\bar{a}$.*

See the article of Ziegler (5) for alternate definitions of one-based.

For example, the theory of algebraically closed fields is not one-based: in general, you cannot recover the coefficients of the equation of a curve from just one point of this curve.

The aim of this section is to prove the following theorem.

Theorem 6.2 *Let G be a one-based group. Then*

1. *G is abelian-by-finite (which means that G has a definable abelian subgroup of finite index)*

2. *If H is a connected definable subgroup of G, then the canonical parameter of H is algebraic over the empty set .*

3. *For any $n \in \omega$, for any $p \in S_n(G)$, there exists $b \in G$ such that*

$$Stab_p [v_0] b \in p.$$

4. *Any definable subset of G^n is a (finite) boolean combination of cosets of definable subgroups of G^n.*

5. *Any definable subset of G^n is a (finite) boolean combination of cosets of definable connected subgroups of G^n.*

Proof: We begin by proving 2. Let g be a realization of the generic of G, G' a big saturated model containing $G \cup \{g\}$, p be the generic of H, p' the heir of p over G' and a a realization of p'. Let $q = t(ga/G')$. Let u be the canonical parameter of H and v the canonical parameter of q.

We first remark that, by one-basedness, v is algebraic over ga. Secondly that $u \in G$; thirdly that $t(ga/G)$ is a generic of G, so ga and u are independent over the empty set; fourthly that $MR(q) = MR(a/G') = MR(H)$.

Last, but a little more difficult, we see that u is definable over v. This will finish the proof of 2, because then we will know that u is algebraic over ga and independent of ga over the empty set, hence is algebraic over the empty set. To do this, it suffices to prove that any automorphism of G' leaving q fixed leaves $H[G']$ setwise fixed.

So let f be an automorphism of G' such that $f(q) = q$, and set $H_1 = f[H]$ and $g_1 = f(g)$. Because the sets gH and g_1H_1 both belong to q, we see that

$$MR(q) = MR(H) \geq MR(gH \cap g_1H_1) \geq MR(q).$$

Thus $MR(gH \cap g_1H_1) = MR(H)$. But it is immediate that $gH \cap g_1H_1 = g_2(H \cap H_1)$, for any $g_2 \in gH \cap g_1H_1$, so $MR(H \cap H_1) = MR(H)$. Since H is connected, this implies that $H_1 = H$, and since $gH \cap g_1H_1$ is not empty this implies that $gH = g_1H_1$.

Let us now prove 1. To do this, we will prove that the connected component of G is abelian. In other words, we will assume that G is connected and prove it is abelian. We will use the above result in G^2. Consider, for any $g \in G$, the following subgroup of G^2 : $H_g = \{(h, g^{-1}hg) \ ; \ h \in G\}$, and the equivalence relation $g \sim g'$ if and only if $H_g = H_{g'}$. We first remark that H_g is isomorphic to G (via the isomorphism $h \rightsquigarrow (h, g^{-1}hg)$). So H_g is connected. We notice also that $g \sim g'$ if and only if, for all $h \in G$, $g^{-1}hg = g'^{-1}hg'$, that is, if and only if g and g' are equivalent mod $Z(G)$ the centralizer of G. It follows that this equivalence relation is definable. Now, since for all $g \in G$, H_g is definable with algebraic parameters and connected, there are, in any elementary extension of G, no more than a countable number of distinct groups of the form H_g, thus the equivalence relation $\sim$ has no more than a countable number of classes, and this is possible only if has a finite number of classes. The conclusion is that $Z(G)$ has finite index in G.

3. We may assume that $n = 1$ and that the canonical base of p is empty (if not, add the necessary parameters as constants to the language). Let g be a realization of the generic type of G, G' a big saturated model containing $G \cup \{g\}$, p' the heir of p over G' and a a realization of p'. Let $q = t(ga/G')$. Since g is generic and g and a are independent over G, a and ga are independent over G. Let S be the stabilizer of p (or of p'), u the canonical parameter of the definable set gS and v the canonical base of the type q.

We first show that u is definable over v and conversely. Since gS is definable with parameters in G' and G' is saturated, it suffices to prove that, for any automorphism f of G', f leaves gS setwise fixed if and only if it leaves the type q fixed. But $q = g \cdot p'$, so $f(q) = f(g) \cdot f(p') = f(g) \cdot p'$ (because the canonical base of p' is empty), thus $f(q) = q$ if and only if $g \cdot p' = f(g) \cdot p'$ if and only if $g^{-1}f(g) \in S$, if and only if $gS = f(g)S$. But $f(S) = S$ (because S is definable without parameter, since the canonical base of p is empty). It follows that $f(q) = q$ if and only if $f(gS) = gS$.

Now, since G is one-based, v is algebraic over ga, and so is u. So, a and $ga \cup \{u\}$ are independent over G, and a and ga are independent over $G \cup \{u\}$. Let a' be an element of G' realizing p and independent with u over G. Since the canonical base of q $(= t(ga/G))$ is definable over u, G' and ga are independent over $G \cup \{u\}$, a' and ga are independent over $G \cup \{u\}$, and a' and $ga \cup \{u\}$ are independent over G. We can conclude that

$$t((ga, u, a)/G) = t((ga, u, a')/G).$$

It is clear that $ga \in gSa$, and that this fact is expressed by a formula with parameters ga, u (to define the set gS) and a. It follows that $ga \in gSa'$ and $a \in Sa'$. Again, this fact can be expressed by a formula with parameters a and a' (S is definable without parameters), so is true for any two independent realizations of p. Thus, if b is a realization of p in G, we have $a \in Sb$, and $Stab_p[v_0]\, b \in p$.

4. To prove that any formula is a boolean combination of formulas in a family $\mathcal{F}$, it suffices to prove that for any two distinct complete types, there is a formula in $\mathcal{F}$ which belongs to the first but not to the second. Thus, we will be home if we prove:

Claim. *Let p and p' be types in $S_n(G)$, and assume that for any definable subgroup H of G, for any $a \in G$, $p \in Ha$ if and only if $p' \in Ha$. Then $p = p'$.*

We may assume that $MR(p') \geq MR(p)$. Let c and c' realize respectively p and p', c and c' independent over G. let S be the stabilizers of p. We know that there exists $a \in G$ such that $c \in Sa$. By hypothesis, it follows that $c' \in Sa$ and $c'c^{-1} \in S$. Notice that $c'c^{-1}$ and c are independent over G. We have:

$$
\begin{array}{rcll}
MR(c'c^{-1}/G \cup \{c\}) & = & MR(c'/G \cup \{c\}) & \text{by } 1.5 \\
 & = & MR(p') & c \text{ and } c' \text{ are independant} \\
 & \geq & MR(p) & \text{by hypothesis} \\
 & \geq & MR(S) & \text{by } 3.3 \\
 & \geq & MR(c'c^{-1}/G) & \text{because } c'c^{-1} \in S
\end{array}
$$

It follows that $p' = t(c'/G) = t(c'c^{-1}c/G) = c'c^{-1} \cdot p = p$.

5. This follows immediately from 4 since if H is a definable subgroup, any class modulo H is a finite union of classes modulo the connected component of H. □

7 Almost strongly minimal subgroups

In this section, G will be a commutative ω-stable group of finite Morley rank. In fact, the results are still valid without the commutativity assumption. This assumption is used only at the end of the section (a product of subgroups is a subgroup), and could be avoided by a more sophisticated argument.

Proposition 7.1 *Let X be a strongly minimal subset of G. Then there exists a connected subgroup H of G such that $H \subseteq acl(X)$ and such that X/H is finite.*

Proof: We noted at the beginning of section 5 that there exists an indecomposable set X_0 included in X such that $X - X_0$ is finite. Let a be an element of X_0 and set $X_1 = a^{-1}X_0$. Then X_1 is indecomposable, contains the unit of G and is contained in the algebraic closure of X. Let H be the subgroup generated by X. It is definable and connected by theorem 5.2. It is clearly in the algebraic closure of X and, since it contains X_1, X/H has only a finite number of elements. □

Let X be a strongly minimal subset of G. Consider the following class of subgroups of G:

$$\mathcal{B}_X = \{B \ ; \ B \text{ is a connected subgroup of } G \text{ and there exists a finite set } F \text{ such that } B \subset acl(F \cup X)\}$$

Since the rank of G is finite, there is at least one element of maximal rank in this class , which in fact is maximal. Moreover, if B_1 and B_2 are two elements of $\mathcal{B}_X$, then B_1B_2, the subgroup generated by $B_1 \cup B_2$ is again in $\mathcal{B}_X$. Consequently, there is a unique maximal element in $\mathcal{B}_X$ which we shall call B_X. By the previous proposition, X/B_X is a finite set.

Moreover, if X and X' are two strongly minimal non orthogonal subsets of G, then there exists a finite set such that $X \subseteq acl(X')$, and conversely, thus $B_X = B_{X'}$.

Consider now the following class, denoted $\mathcal{B}$:

$$\{B \ ; \ B \text{ is a connected subgroup of } G \text{ and there are } n \in \omega, \text{ strongly minimal sets } X_1, X_2, \ldots, X_n \text{ and a finite set } F \text{ such that } B \subset acl(F \cup \textstyle\bigcup_{1 \leq i \leq n} X_i)\}$$

As above, we see that $\mathcal{B}$ has a unique maximal element. Call B this maximal element and F a finite subset such that $B \subset acl(F \cup \bigcup_{1 \leq i \leq n} X_i)$. It is clear that B contains B_X for any strongly minimal set X.

Proposition 7.2 *There exist an integer n and strongly minimal sets X_1, X_2, $\ldots, X_n$ such that $B = B_{X_1} B_{X_2} \ldots B_{X_n}$. Moreover we may assume that the X_i's for $1 \leq i \leq n$ are pairwise orthogonal.*

Proof: Consider once more the class of subgroups of G of the form $B_{Y_1} B_{Y_2} \ldots B_{Y_n}$, where the Y_i are strongly minimal. By the indecomposability theorem 5.2, they are connected, and, as above, there exists a unique maximal subgroup in this class. Let $H = B_{X_1} B_{X_2} \ldots B_{X_n}$ be this maximal subgroup. We remark that $H \subseteq B$, and that for any strongly minimal set X, $B_X \subseteq H$, and consequently, X/H is finite. This last property extends to any rank 1 set, since such a set can be split into a finite number of strongly minimal sets.

We want to show that $H = B$, and so, we assume, toward a contradiction, that $H \neq B$. It follows that B/H is infinite, so there exists an element $c \in B$ such that the class $\widetilde{c} = cH$ is not algebraic over F. On the other hand, there exists a finite set $Y \subseteq \bigcup_{1 \leq i \leq = n} X_i$ such that $c \in acl(F \cup Y)$. We choose a minimal such Y. It is clear that $\widetilde{c} \in acl(F \cup Y)$, so there exist $Y' \subseteq Y$ and $y \in Y$ such that $\widetilde{c} \in acl(F \cup Y' \cup \{y\})$, but $\widetilde{c} \notin acl(F \cup Y')$. By minimality of $t(y/F)$, it follows that $y \in acl(F \cup Y' \cup \{\widetilde{c}\})$. Set $Y_0 = Y - \{y\}$. Thus $c \notin acl(F \cup Y_0)$ (by minimality of Y), $c \in acl(F \cup Y_0 \cup \{y\})$ and $c \in acl(F \cup Y_0 \cup \{\widetilde{c}\})$.

This implies in particular that $RM(c/F \cup Y_0) = 1$, that is, c is contained in a rank 1 set T definable with parameters in $F \cup Y_0$. Now, since $c \in acl(F \cup Y_0 \cup \{\widetilde{c}\})$, there exists a formula $\phi(x, y)$ with parameters in $F \cup Y_0$ and an integer n such that

$$G \models \phi(c, \widetilde{c}) \wedge \exists^{\leq n} x \phi = (x, \widetilde{c}).$$

So c satisfies the formula $\phi(x, \widetilde{x}) \wedge \exists^{\leq n} x \phi(x, \widetilde{x})$ and we may assume that every element of T satisfies this formula (replacing T by $T \wedge \phi(x, \widetilde{x}) \wedge \exists^{\leq n} x \phi(x, \widetilde{x})$). Thus, for any $d \in T$, d is algebraic over $F \cup Y_0 \cup \left\{\widetilde{d}\right\}$. But we have already remarked that $\left\{\widetilde{d}\ ;\ d \in T\right\}$ is finite, and this is a contradiction with the fact that T itself is infinite.

The reason we may assume that the X_i's for $1 \leq i \leq n$ are pairwise orthogonal is that, as we noted before, if X_i and X_j are non orthogonal, then $B_{X_i} = B_{X_j}$. □

For more information about stable groups, one can consult one of the following books:

1. [Po 87]: in french, a good introduction to model theoretic methods in the subject, accessible to non specialists of model theory.

2. [Pi2 96]: the most recent book on the subject of geometric stability theory.

3. [NePi]: a multi-authors collection of articles intended as an introduction to stable groups also accessible to non specialists.

4. [BoNe]: an exhaustive account of the relations between groups of finite Morley rank, algebraic groups and finite groups.

References

[BoNe] A. Borovik and A. Nesin, *Groups of finite Morley Rank*, Oxford Logic Guides 26, Oxford Science publications, 1994.

[NePi] A. Nesin and A. Pillay ed., *The Model Theory of Groups*, Notre Dame University Press, 1989.

[Pi2 96] A. Pillay, *Geometrical Stability Theory*, Oxford University Press, 1996

[Pi1] A. Pillay, *Model theory of algebraically closed fields*, this volume.

[Po 87] B. Poizat, *Groupes Stables*, Nur al-matiq wal ma'rifah, Villeurbanne, France, 1987.

[Zie] M. Ziegler, *Introduction to stability theory and Morley rank*, this volume.

Model theory of algebraically closed fields

ANAND PILLAY

We give a survey of the model-theoretic approach to algebraically closed fields, algebraic varieties and algebraic groups. Much of what we say is taken quite directly from other sources, specifically [Po 89],[Po 87, Chapter 4], [Bousl 89], and [Pi 89], as well as from basic textbooks on algebraic geometry and algebraic groups ([Sh],[Bor]). As we tend to be brief with our proofs, the reader is advised to look at these other sources for additional details, where appropriate. Also all relevant attributions of results can be found there. The reader should see [Zie] in this volume for ω-stability, imaginaries, canonical bases etc. The present paper can serve as an introduction to naive algebraic geometry for model-theorists, as all the basic notions will be defined.

1 Algebraically closed fields

A field F is said to be algebraically closed if whenever $P(X) = X^n + a_{n-1}X_{n-1} + \ldots + a_1X + a_0$ is a polynomial over F in the single indeterminate X, and $n > 0$, then $P(X) = 0$ has a solution in F. The Fundamental Theorem of Algebra states that the field of complex numbers is algebraically closed. On the other hand the existence of algebraically closed fields in all positive characteristics (in fact the existence of the "algebraic closure" of any given field) is a classical fact (see Chapter VII of [Lan]).

Let L be the language consisting of binary operations $+, -, .$, and constant symbols $0, 1$. Fix p to be either 0, or a prime number. Then we can clearly "write down" a set of L-sentences expressing that F is an algebraically closed field of characteristic p. We call the resulting theory ACF_p. Note that an L-substructure of a field F is in general a subring of F (not a subfield).

Proposition 1.1 *ACF_p is complete and has quantifier elimination.*

Proof: Let F_1, F_2 be ω-saturated models of ACF_p. We will show that the set I of partial isomorphisms between finitely generated substructures of F_1 and F_2, is nonempty and has the back-and-forth property. (This will guarantee that if $\bar{a} \in F_1, \bar{b} \in F_2$ are (possibly empty) sequences with the same quantifier-free type, then $(F_1, \bar{a}) \equiv (F_2, \bar{b})$.) This shows first that $F_1 \equiv F_2$, whereby ACF_p is complete, and secondly that if F is a model of ACF_p then the type of a

Author partially supported by a grant from the NSF.

tuple from F is determined by its quantifier-free type. The latter, together with compactness, yields that ACF_p has quantifier elimination.

(i) **I is nonempty:**

The substructure of F_1 generated by $\emptyset$ (i.e. by the constants 0,1) is either Z (if $p = 0$), or the finite field F_p (if $p > 0$). Similarly for F_2.

(ii) (back-and-forth).

Let $f : R_1 \cong R_2$ be an isomorphism between the finitely generated substructures R_1, R_2 of F_1, F_2. Clearly f extends to an isomorphism (which we call f again) between the quotient fields L_1, L_2 of R_1, R_2. Let $\alpha \in F_1$.

CASE (a). α is algebraic over L_1 (namely is the zero of some nonzero polynomial in $L_1[X]$). Let $P(X) \in L_1[X]$ be the minimal polynomial of α over L_1. P is irreducible and generates the ideal $I(\alpha/L_1)$ of all polynomials over L_1 vanishing at α. (We use here the fact that $L_1[X]$ is Euclidean.)

Let $Q(X) \in L_2[X]$ be the image of P under the isomorphism $f : L_1 \cong L_2$. As F_2 is algebraically closed, $Q = 0$ has a solution β in F_2; f then extends to an isomorphism g: $L_1(\alpha) \cong L_2(\beta)$ by putting $g(\alpha) = \beta$.

CASE (b). α is transcendental over L_1 (namely is not algebraic over L_1). Now F_2 is infinite and L_2 is finitely generated. Every nonzero polynomial over L_2 has at most finitely many solutions in F_2. Thus by ω-saturation of F_2, there is $\beta \in F_2$ such that β is transcendental over L_2. Clearly putting $g(\alpha) = \beta$ yields an extension $g : L_1(\alpha) \cong L_2(\beta)$ of f. □

Corollary 1.2 *ACF_p is strongly minimal.*

Proof : Let F be an algebraically closed field, and let $X \subseteq F$ be a definable set. By 1.1, X is a finite Boolean combination of zero sets of polynomials $P(X) \in F[X]$. But any such zero set is finite (or all of F, if P is 0), so X is finite or cofinite. □

Corollary 1.3 *Let F be an algebraically closed field. Let $A \subseteq F, a \in F$, and let $k < F$ be the field generated by A. Suppose a is algebraic over A in the model theoretic sense. Then a is algebraic over k in the field theoretic sense. Moreover $tp(a/A)$ is isolated by the formula "$P(x) = 0$" where $P(X)$ is the minimal polynomial of a over k.*

Proof: We may suppose F to be $|A|^+$-saturated. If a were transcendental over k, then there would be (by saturation of F) infinitely many $b \in F$ which are also transcendental over k and thus (by 1.1) have the same type as a over k. This would contradict the assumption that a is model-theoretically algebraic over k. This proves the first part.

Let $P(X)$ be the minimal polynomial of a over k. The formula $P(x) = 0$ is clearly equivalent to a formula over A. Also P is irreducible. If also $P(b) = 0$, then $I(a/k) = I(b/k)$, so by 1.1, $tp(a/k) = tp(b/k)$, so $tp(a/A) = tp(b/A)$. □

Note that (with the notation of 1.3), if a is field-theoretically algebraic over k, then a is model-theoretically algebraic over A. So the expression $a \in acl(A)$ can, and will be, used unambiguously.

If k is a field of characteristic $p > 0$, then the map $x \to x^p$ is a field isomorphism of k with its image. This map is called the Frobenius map, denoted Fr. The field k is said to be perfect if $Fr : k \to k$ is surjective. (If k has characteristic $0, k$ is always said to be perfect.) If k is a perfect field, then any irreducible polynomial $P(X)$ over k has no multiple roots (see [Lan]). Suppose that F is algebraically closed, and $k < F$. By k_{ins} we mean $\cup_n Fr^{-n}(k)$. So k_{ins} is the smallest perfect field containing k.

Corollary 1.4 *Suppose F is algebraically closed, $A \subseteq F$ and $k < F$ is the field generated by A. Let $a \in F$. Then $a \in dcl(A)$ iff $a \in k_{ins}$.*

Proof: Right to left is clear.

Left to right: Note that $a \in dcl(k_{ins})$. By 1.3, $tp(a/k_{ins})$ is isolated by "$P(x) = 0$" where $P(X)$ is the minimal polynomial of a over k_{ins}. If $P(X)$ is of degree > 1, then by above remarks, $P(X) = 0$ has more than one solution, contradicting $a \in dcl(k_{ins})$. Thus $P(X)$ has degree 1, namely $a \in k_{ins}$. □

Corollary 1.5 *Suppose F is an algebraically closed field. Let k be a perfect subfield, $X \subseteq F^n$ some k-definable set, and $f : X \to F$ some k-definable function. Then X can be written as a finite union of k-definable sets $X_1 \cup \ldots \cup X_m$ such that for each i, there is some k-rational function $f_i(x)$ and some $j(i) \leq 0$ such that $f|X_i = (Fr^{j(i)} \cdot f_i)|X_i$.*

Proof: By Corollary 1.4 and compactness. □

Our next aim is to show that algebraically closed fields have "elimination of imaginaries". Recall (see [Zie, 1.5]) that a structure M is said to have elimination of imaginaries, if for any $e \in M^{eq}$, there is some tuple $\bar{b}$ from M such that $e \in dcl(\bar{b})$ and $\bar{b} \in dcl(e)$. M is said to have weak elimination of imaginaries if for any $e \in M^{eq}$ there is some tuple $\bar{b}$ from M such that $e \in dcl(\bar{b})$ and $\bar{b} \in acl(e)$. The complete theory T is said to have elimination of imaginaries if every model (or equivalently some reasonably saturated model) of T has the property. Similarly for weak elimination of imaginaries.

Lemma 1.6 *Let T be a strongly minimal theory with the feature that for any (or equivalently some) model M of $T, acl(\emptyset) \cap M$ is infinite. Then T has weak elimination of imaginaries.*

Proof: Let M be some saturated model of T. Let $e \in M^{eq}$. Let $\bar{b} = (b_1, \ldots, b_n)$ be some tuple from M such that $e \in dcl(\bar{b})$. Let f be a $\emptyset$-definable function such that $f(\bar{b}) = e$. Let $M_0 = acl(e) \cap M$. By our assumptions M_0 is infinite. We will find $c_1, \ldots, c_n$ in M_0 such that $e = f(\bar{c})$. The c_i are found inductively. Suppose, by induction hypothesis that we already have $c_1, \ldots, c_{i-1}$ in M_0 such that for some $x_i, \ldots, x_n$ in M, $e = f(c_1, \ldots, c_{i-1}, x_i, \ldots, x_n)$. Let $\varphi(x_i)$ be the formula $(\exists x_{i+1}, \ldots, x_n)(e = f(c_1, \ldots, c_{i-1}, x_i, x_{i+1}, \ldots, x_n))$. Then either $\varphi(x_i)$ has only finitely many solutions in M, which must therefore be all contained in $acl(e, c_1, \ldots, c_{i-1}) \cap M = M_0$ or $\varphi(x_i)$ has cofinitely many solutions in M, whereby, as M_0 is infinite, some such solution is in M_0. Either way $\varphi(x_i)$ is realized in M_0.

Thus we can find a tuple $\bar{c}$ from M_0 such that $e = f(\bar{c})$. So $e \in dcl(\bar{c})$ and $\bar{c} \in acl(e)$. □

Note that ACF_p satisfies the hypothesis of Lemma 1.6, so ACF_p will have weak elimination of imaginaries. To deduce full elimination of imaginaries we require:

Lemma 1.7 *Let F be an algebraically closed field. Let $\bar{c}_1, \ldots, \bar{c}_m$ be a finite set of n-tuples from F. Let e be the finite set $\{\bar{c}_1, \ldots, \bar{c}_m\}$ (thought of as an element of F^{eq}). Then there is some tuple $\bar{d}$ from F such that $e \in dcl(\bar{d})$ and $\bar{d} \in dcl(e)$. (Assuming F to have infinite transcendence degree, the condition on $\bar{d}$ is equivalent to: for any automorphism f of F, f fixes $\bar{d}$ iff f permutes $\{\bar{c}_1, \ldots, \bar{c}_m\}$.)*

Proof : Let $Z, X_1, \ldots, X_n$ be indeterminates. Write $\bar{c}_i$ as $(c_{i1}, \ldots, c_{in})$. Let $P(Z, X_1, \ldots, X_n)$ be the polynomial:

$$(Z+c_{11}X_1+\ldots+c_{1n}X_n)(Z+c_{21}X_1+\ldots+c_{2n}X_n)\ldots\ldots(Z+c_{m1}X_1+\ldots+c_{mn}X_n)$$

Let $\bar{d}$ be the tuple of coefficients of $P(Z, X_1, \ldots, X_n)$. As $F[Z, X_1, \ldots, X_n]$ is a unique factorization domain, it is clear that $\bar{d}$ works. □

REMARK From the model-theoretic point of view, the manner in which the tuple $\bar{d}$ was obtained is not particularly relevant. However it is rather important for the geometric view. The tuple $\bar{d}$ given by the proof above is called the tuple of <u>Chow coordinates</u> of $\{\bar{c}_1, \ldots, \bar{c}_m\}$.

Corollary 1.8 *ACF_p has elimination of imaginaries.*

Proof : Let $F \models ACF_p$. Let $e \in F^{eq}$. Let $\bar{c}$ be a tuple from F given by Lemma 1.6 (namely $e \in dcl(\bar{c})$ and $\bar{c} \in acl(e)$). Let $\{\bar{c}_1, \ldots, \bar{c}_m\}$ be the set of e-conjugates of $\bar{c}$ (namely tuples realizing the same type over e as $\bar{c}$). Let e' be this finite set as an element of F^{eq}. Clearly $e \in dcl(e')$ and $e' \in dcl(e)$. Now apply Lemma 1.7 to e', to conclude. □

Example 1.9 *Let F be an algebraically closed field, and let G be a group definable in F, in the sense that both G and the graph of the group operation of G are definable subsets (of F^n, F^{3n} respectively). Let H be a definable subgroup of G. Then there is some set X definable in F (so a definable subset of F^m for some m), and a definable bijection between the homogeneous space G/H and X.*

We recall now some facts about strongly minimal sets (see in [Pi 89] 2.4, [Pi2 96], or [Zie] Section 5 in this volume for more details). RM denotes Morley rank, $mult(X)$ denotes the Morley multiplicity or degree of X and $\phi(x,b)^M$ denotes the set of solutions of the formula $\phi(x,b)$ in the model M.

Definition/Facts 1.10 *Let M be a saturated strongly minimal structure.*
(i) Let a be a tuple from M and $A \subseteq M$. Let $(b_1, \ldots, b_k)$ be a maximal subtuple of a such that $\{b_1, \ldots, b_k\}$ is algebraically independent over A (namely $b_i \notin acl(A \cup \{b_j : j \neq i\})$. Then k depends only on a, and we define $dim(a/A)$ to be k.
(ii) Let $X \subseteq M^n$ be A-definable, where $A \subseteq M$ is small (namely of cardinality less than the cardinality of M). By $dim(X)$ we mean $max\{dim(a/A) : a \in X\}$. This definition does not depend on A.
(iii) For any formula $\varphi(x, y)$ of the language of M, and $k < \omega$, there is a formula $\Psi(y)$ such that for any b from M, $M \models \Psi(b)$ iff $dim(\varphi(x, b)^M) = k$.
(iv) For any definable set $X (\subseteq M^n)$, $dim\ (X) = RM(X)$.
(v) If $X \subseteq M^n$ is A-definable, then $a \in X$ is said to be a generic point of X over A if $dim\ (X) = dim\ (a/A)$, and moreover $tp(a/A)$ is said to be a generic type of X over A.
(vi) For each n, M^n has multiplicity (or Morley degree) 1. (There is, for $A \subseteq M$, a unique type over A of n algebraically independent over A elements of M.)
(vii) If F is a model of $ACF_p, k < F$ and $a \in F^n$, then $dim\ (a/k) = tr.degree\ (k(a)/k)$.

We want to explain how Morley rank is determined by Zariski closed sets in the algebraically closed field context.

Definition 1.11 *Let F be a field. By a Zariski closed subset of F^n we mean a set X which is a finite intersection of zero sets of polynomials over F (in n indeterminates).*

(Clearly such a set is definable in F.)

Lemma 1.12 *Let $F \models ACF_p$. Let $A \subseteq F$, and $\bar{a} \in F^n$. Suppose $RM(tp(\bar{a}/A)) = m$. Then there is an A-definable Zariski closed set X such that the formula "$x \in X$" isolates $tp(\bar{a}/A)$ among complete types over A of Morley rank $\leq m$ (in other words, $tp(\bar{a}/A)$ is the unique generic type of X over A).*

Proof : First, replacing A by the field it generates, we may assume that A is a subfield k of F.

Claim. Let $\bar{b} = (b_1, \ldots, b_r)$ be any tuple from F. Suppose that $s \leq r$, and $\dim\ (b_1, \ldots, b_s/k) = \dim\ (b_1, \ldots, b_r/k) = s$. Then there is a finite set $\bar{P}$ of polynomial equations over k in indeterminates $X_1, \ldots, X_r$ such that whenever $\bar{c} \in F^r, P(\bar{c}) = 0$ for all $P \in \bar{P}$, and $\dim\ (c_1, \ldots, c_s/k) = s$, then $tp(\bar{c}/k) = tp(\bar{b}/k)$.

Proof of claim. Iterating Corollary 1.3, we may find a finite set $\bar{Q}$ of polynomials over (the field) $k(b_1, \ldots, b_s)$ in indeterminates $X_{s+1}, \ldots, X_r$, such that the formula $\bigwedge\{Q(x_{s+1}, \ldots, x_r) = 0 : Q \in \bar{Q}\}$ isolates

$$tp(b_{s+1}, \ldots, b_r/k(b_1, \ldots, b_s)).$$

The coefficients of each polynomial $Q \in Q$ are k-rational functions of $b_1, \ldots, b_s$, the denominators of which are nonzero. Multiplying through by the denominators of each such coefficient, and replacing $b_1, \ldots, b_s$ by indeterminates $X_1, \ldots, X_s$, yields a finite set $\bar{P}$ of polynomials over k in $X_1, \ldots, X_r$. It should be clear that $\bar{P}$ works. (Suppose $(c_1, \ldots, c_r) \in F^r$ and dim $(c_1, \ldots, c_s/k) = s$ and $P(\bar{c}) = 0$ for all $P \in \bar{P}$. In particular $tp(c_1, \ldots, c_s/k) = tp(b_1, \ldots, b_s/k)$. So we may assume that $c_i = b_i$ for $i = 1, \ldots, s$. Dividing by the same polynomials that we multiplied the $Q \in \bar{Q}$ by earlier, shows that $(c_{s+1}, \ldots, c_r)$ is a common zero of the set $\bar{Q}$. By choice of $\bar{Q}$, $tp(c_{s+1}, \ldots, c_r/k(b_1, \ldots, b_s)) = tp(b_{s+1}, \ldots, b_r/k(b_1, \ldots, b_s))$, and so $tp(\bar{c}/k) = tp(\bar{b}/k)$.) This proves the claim.

We now return to the tuple $\bar{a}$ mentioned in the statement of the Lemma. For each subtuple $(a_{i(1)}, \ldots, a_{i(r)})$ of $\bar{a}$ which satisfies the hypothesis of the claim, let $\bar{P}_{(i(1), \ldots, i(r))}$ be a finite set of polynomials in $X_{i(1)}, \ldots, X_{i(r)}$, given by the claim. Let $\bar{P}$ be the union of these finite sets of polynomials. We show that (the Zariski closed set defined by) $\bigwedge\{P(X) = 0 : P \in \bar{P}\}$ works. So let $\bar{b} = (b_1, \ldots, b_n)$ be a common zero of the set $\bar{P}$ with $\dim(\bar{b}/k) \geq m$. We must show that $tp(\bar{b}/k) = tp(\bar{a}/k)$. Let $1 \leq i(1), \ldots, i(m) \leq n$ be such that $\dim(b_{i(1)}, \ldots, b_{i(m)}/k) = m$. By considering the polynomials in $\bar{P}_{(i(1))}, \ldots, \bar{P}_{(i(1), \ldots, i(m))}$ successively, we see that actually dim $(a_{i(1)}, \ldots, a_{i(m)}/k) = m$ too. (If $a_{i(1)}$ were algebraic over k, then as $b_{i(1)}$ satisfies the polynomials in $\bar{P}_{i(1)}$, $b_{i(1)}$ would also be algebraic over k, which it is not. Continue this way.) As dim $(\bar{a}/k) = m$, we see, by considering $\bar{P}_{i(1), \ldots, i(m), \ldots, i(n)}$ that $tp(\bar{b}/k) = tp(\bar{a}/k)$. The lemma is proved. □

Finally in this section we point out:

Lemma 1.13 *ACF_p is not locally modular.*

Proof: Let F be a saturated algebraically closed field. For $c, d \in F$, the formula $y = cx + d$ clearly defines a strongly minimal set $L_{c,d} \in F^2$ (namely a line). Choose c, d algebraically independent over $\emptyset$. Let $p_{c,d}(x, y) \in S(F)$ be the generic type of $L_{c,d}$. So $RM(p_{c,d}) = 1$. We claim that $Cb(p_{c,d}) = (c, d)$. For if f is an automorphism of F which fixes $p_{c,d}$ then $f(L_{c,d}) = L_{f(c), f(d)}$ must meet $L_{c,d}$ in an infinite set. This is only possible if $f(c) = c$, and $f(d) = d$. So $RM(Cb(p_{c,d})) = 2$. This yields non local modularity [Zie, 5.12]. □

2 Zariski closed sets

Recall:

Fact 2.1 ([Lan]) *If F is a field, then for all n, the ring $F[X_1, \ldots, X_n]$ is Noetherian, namely satisfies the ACC on ideals or equivalently satisfies: every ideal is finitely generated.*

It follows from 2.1 that

Fact 2.2 *If F is a field, then the intersection of any family of Zariski closed subsets of F^n is equal to a finite subintersection. In other words we have the DCC on Zariski closed sets. Also note that any finite union of Zariski closed subsets of F^n is Zariski closed.*

Lemma 2.3 (Hilbert's Nullstellensatz) *Let F be an algebraically closed field, and $I \subseteq F[X_1, \ldots, X_n]$ an ideal, and $f \in F[X_1, \ldots, X_n]$. Suppose that for all $\bar{a} \in F^n$, whenever $P(\bar{a}) = 0$ for all $P \in I$, then $f(\bar{a}) = 0$. Then for some k, $f^k \in I$.*

Proof: We may assume $I \neq F[X_1, \ldots, X_n]$. If $f^k \notin I$ for all k, let J be maximal among ideals in $F[X_1, \ldots, X_n]$ which include I, but do not contain f^k for all k. It is easy to prove that J is prime. Let $R = F[X]/J$, an integral domain, and let L be its quotient field. Then L contains F. Let $\bar{b}$ be the image of $\bar{X}$ in L. Let $P_1, \ldots, P_r$ be generators of J. Then $P_i(\bar{b}) = 0$ for $i = 1, \ldots, r$, but $f(\bar{b}) \neq 0$. As $Th(F)$ has quantifier elimination, F is an elementary substructure of the algebraic closure of L. Thus there is $\bar{a} \in F^n$ such $P_i(\bar{a}) = 0$ for $i = 1, \ldots, r$, but $f(\bar{a}) \neq 0$. As the P_i generate J and $J \supseteq I$ we have $P(\bar{a}) = 0$ for all $P \in I$, a contradiction. □

Corollary 2.4 *Let F be an algebraically closed field. Then there is a bijection between Zariski closed subsets of F^n and radical ideals $I \subseteq F[X_1, \ldots, X_n]$, given by: if $V \subseteq F^n$ is Zariski closed then $I(V) = \{P(\bar{X}) \in F[\bar{X}] : P(\bar{a}) = 0$ for all $\bar{a} \in V\}$, and if $I \subseteq F[\bar{X}]$ is radical then $V(I)$ is simply the zero set of I in F^n.*

Let us now fix a big algebraically closed field K (a universal domain both for algebraic geometry and model theory). By the DCC on Zariski closed sets it follows formally that any Zariski closed (or as we shall just say, closed) subset V of F^n can be uniquely written as a nonredundant finite union of irreducible Zariski closed sets (where a Zariski closed set is said to be irreducible if it can not be written as a union of two proper Zariski closed subsets). Moreover the correspondence between closed sets and radical ideals shows that V is irreducible iff $I(V)$ is a prime ideal of $K[X]$.

Algebraic geometry has a notion of "defined over" which, in positive characteristic, has some discrepancy with the model-theoretic notion:

Definition 2.5 *Let $V \subseteq K^n$ be Zariski closed, and let $k < K$. V is said to be defined over k (in the field-theoretic sense) if $I(V)$ is generated by polynomials in $k[X_1, \ldots, X_n]$.*

Recall on the other hand that V is said to be defined over k in the model-theoretic sense just if V can be defined by some formula $\varphi(\bar{x})$ with parameters in k (where $\varphi(\bar{x})$ does not, a priori, even have to be a conjunction of polynomial equations).

Fact 2.6 *Let $V \subseteq K^n$ be Zariski closed. Then V has a smallest field of definition (in the field theoretic sense). If k is such, then it has the feature that any automorphism of K fixes V setwise iff it fixes k pointwise.*

Proof: (Sketch) Let $I = I(V)$. Then $R = K[X]/I$ is a vector space over K, and we may choose a set M of monomials, which mod I forms a basis for R. Any monomial can be written uniquely as a K-linear combination of monomials in

M (mod I). Let k be the field generated by all the coefficients arising in this fashion. □

Fact 2.6 provides another route to (weak) elimination of imaginaries in ACF, but here we just point out how it relates the two notions of "defined over".

Corollary 2.7 *Let $V \subseteq K^n$ be Zariski closed. Let $k < K$. Then V is defined over k in the model-theoretic sense iff V is defined over k_{ins} in the field-theoretic sense. (In particular the notions agree if k is perfect.)*

Proof : Right to left is immediate. For left to right: Let k_0 be the smallest field of definition of V given by 2.6. Suppose that V is defined over k in the model-theoretic sense. Then (by 2.6) $k_0 \subseteq dcl(k)$. By Corollary 1.4, $k_0 \subseteq k_{ins}$. Thus V is defined over k_{ins} in the field-theoretic sense. □

In some parts of the literature one finds the notion of a Zariski closed set V being k-closed: meaning that V is the zero set of a finite set of polynomials in $k[\bar{X}]$. From the above remarks it should then be clear that V is k-closed iff V is defined over k in the model-theoretic sense. For our purposes we may usually assume that any small subfield k we deal with is perfect, whereby all the notions: defined over k in model-theoretic sense, defined over k in field-theoretic sense, and k-closed, are equivalent.

We call V k-irreducible if V is k-closed but it is not the union of two proper k-closed subsets. (Note irreducible means K-irreducible, which is equivalent to k-irreducibility for some algebraically closed subfield k of K.)

Lemma 2.8 *Let $V \subseteq K^n$ be k-closed and k-irreducible. Then V has a unique generic type over k. This type (say $p(\bar{x})$) is characterised by: for k-closed $W \subseteq K^n$, "$\bar{x} \in W$"$\in p$ iff $V \subseteq W$. (In particular, if V is irreducible then $mult(V) = 1$).*

Proof: By compactness, k-irreducibility of V and quantifier elimination, there is a complete type $p(\bar{x})$ over k such that for k-closed W, "$\bar{x} \in W$" $\in p$ iff $V \subseteq W$. Let $RM(p) = m$. By Lemma 1.12, there is k-closed X such that p is isolated among complete types over k of $RM \geq m$ by "$\bar{x} \in X$". But V is the smallest k-closed set in p, so we may assume $V = X$. So $RM(V) = m$, and p is the unique type over k of $RM \geq m$ which contains V. This shows that V has a unique generic type over k. □

Lemma 2.9 *Let V be k-closed and k-irreducible. Then the irreducible components of V are k-conjugate and each is defined over $acl(k)$. (In particular $RM(V) = RM(V_i)$ for any irreducible component V_i of V.)*

Proof : Let $V_1, \ldots, V_r$ be the distinct irreducible components of V. Any k-automorphism of K fixes V setwise so permutes the V_i. This shows that each V_i is (model-theoretically) defined over $acl^{eq}(k) = acl(k)$. Let W_1 be the union of the k-conjugates of V_1, and W_2 the union of the remaining V_i. Then both W_1, W_2 are (model-theoretically) defined over k, hence k-closed, and $V = W_1 \cup W_2$. Hence, by k-irreducibility of V, $W_1 = V$. □

Lemma 2.10 *Let V be closed and irreducible. Then $RM(V) \geq n+1$ iff V contains a proper irreducible closed subset W with $RM(W) \geq n$.*

Proof : First suppose W is a proper irreducible closed subset of V. Let k be such that both V, W are k-closed. By the characterisation in 2.8 of the generic types over k of V and W it follows that $RM(W) < RM(V)$. On the other hand, if $RM(V) = n+1$ say, and V is defined over k, let $\bar{a}$ be a generic point of V over k. so $dim(\bar{a}/k) = n+1$. We may suppose that $a_1 \notin k$. Consider $q(x_2, \ldots) = tp(a_2, \ldots /acl(k \cup \{a_1\}))$. Then $RM(q) = n$, and by 1.12, q is the generic type over $acl(k \cup \{a_1\})$ of some closed (irreducible) W. Then $RM(W) = n$, and $W \subseteq V$. □

Corollary 2.11 *If V is closed and irreducible then $RM(V)$ = Krull dimension of $I(V)$. (Here the Krull dimension of a prime ideal P in a ring R is defined to be the greatest $n < \omega$, if such exists, such that there exists a strictly increasing chain $P_n \supset P_{n-1} \supset \ldots \supset P_0 = P$ of prime ideals of R.)*

Remark 2.12 *For any $\bar{a}, k, mult(tp(\bar{a}/k)) = 1$ iff k_{ins} is algebraically closed in $(k(\bar{a}))_{ins}$.*

Proof: Let V be the smallest k-closed set such that $\bar{a} \in V$. Then $\bar{a}$ is a generic point of V over k, and "$mult(tp(\bar{a}/k)) = 1$" is then clearly equivalent to V being irreducible. Let W be an irreducible component of V containing $\bar{a}$. W is defined over $acl(k)$, and as $RM(V) = RM(W)$), by 2.8 we see that $tp(\bar{a}/acl(k))$ is the unique generic type of W over $acl(k)$. Thus W is the unique irreducible component of V containing $\bar{a}$. Thus, if (by elimination of imaginaries), e is the tuple from K interdefinable with W, we see that $e \in dcl(k(\bar{a}) = (k(a))_{ins}$. But $e \in acl(k)$ (by 2.9). So if k_{ins} is algebraically closed in $(k(\bar{a}))_{ins}$ then W is also k-closed, so $W = V$, and $mult(tp(\bar{a}/k)) = 1$. On the other hand, if there is some $e \in acl(k) \cap dcl(k(\bar{a})), e \notin dcl(k)$, then $mult(tp(e/k) > 1$, so $mult(tp(\bar{a}/k)) > 1$. □

If A is any subset of K^n then by the Zariski closure of A we mean the smallest Zariski closed set $V \subseteq K^n$ such that $A \subseteq V$. (Such V exists by DCC on Zariski closed sets.)

Remark 2.13 *Let $k < K$. Let A be any subset of k^n. Let V be the Zariski closure in K^n of A. Then V is k-closed.*

Proof: Any k-automorphism of K will fix A pointwise and thus fix V setwise. So V is defined over k in the model-theoretic sense, so is k-closed (and moreover defined over k in the field theoretic sense if k is perfect). □

Remark 2.14 *(i) Everything we have done so far can be seen to hold at a very general level, the crucial points being the DCC on closed sets and Lemma 1.12. It turns out that the specificity of algebraically closed fields (among strongly minimal structures with the above features) is captured by the dimension theorem:*

If V, W are closed irreducible subsets of K^n, and U is an irreducible component of $V \cap W$ then $RM(U) \geq RM(V) + RM(W) - n$.
(ii) Our level of analysis is naive set-theoretic. It would be useful to incorporate the sheaf-theoretic point of view and more generally the scheme-theoretic point of view into the model-theoretic approach.

3 Varieties

K remains a big algebraically closed field. We follow the point of view in [We 62], although we have quite possible introduced our own errors and incorrect formulations. The term "affine variety" is sometimes reserved for an irreducible Zariski closed subset of K^n (sometimes equipped with "extra structure"). Here we will simply mean a closed subset of K^n. If V is such then the coordinate ring $K[V]$ of V (namely $K[X_1, \ldots, X_n]/I(V)$) is a finitely generated K-algebra with no nilpotent elements. Moreover any K-algebra R which is finitely generated without nilpotents is of the form $K[V]$ for some affine variety. V is irreducible if and only if $K[V]$ is an integral domain. In this case we can form the quotient field of $K[V]$, which is denoted $K(V)$ and is called the "function field" of V. Note that if V is an affine variety which is defined over k (in the sense of 2.5) then $K[V] = k[V] \otimes_k K$ (where $k[V] = k[X_1, \ldots, X_n]/(I(V) \cap k[X_1, \ldots, X_n])$).

Remark 3.1 *Let V be an irreducible affine variety defined over (algebraically closed) k. Let a be a generic point of V over k. then $k[V] \cong k[a]$, and $k(V) \cong k(a)$ (where the isomorphisms are the natural ones). Thus $RM(V) = tr.degree\ k(V)/k) (= tr.degree\ K(V)/K))$.*

Definition 3.2 *If $V \subseteq K^n$, and $W \subseteq K^m$ are affine varieties, then by a morphism from V to W, we mean a map $f = (f_1, \ldots, f_m) : V \to W$ such that $f_i \in K[V]$ for all i. The map f is said to be an isomorphism if f is $1-1$, onto and f^{-1} is also a morphism from W to V. (So a morphism between affine varieties is simply a polynomial map.)*

Definition 3.3 *By a quasi-affine variety we mean a Zariski open subset of an affine variety. Note that a quasi-affine variety is defined by a finite conjunction of polynomial equations together with a finite disjunction of polynomial inequations. A quasi-affine variety U has its own Zariski topology, namely the relative topology; U will be irreducible just if its Zariski closure is irreducible.*

Definition 3.4 *Let $V \subseteq K^n$ be an affine variety.*
(i) Let $a \in V$. A (not everywhere defined) function f from V to K is said to be regular at a, if there is a Zariski open subset U of V containing a, and there are $P(\bar{X}), Q(\bar{X})$ in $K[X_1, \ldots, X_n]$ such that Q is everywhere nonzero on U, and on $U, f = P/Q$.
(ii) Let U be an open subset of V. A function $f : U \to K$ is said to be regular if it is regular at each point $a \in U$.

(iii) If U_1, U_2 are quasi-affine varieties, then a morphism from U_1 to U_2 is just a map from U_1 to U_2 whose coordinates are regular functions on U_1. As above we also have the notion of an isomorphism.

Remark 3.5 *(i) Note that, by compactness (or even the DCC), if U is quasi-affine and $f : U \to K$ is regular then there is a finite covering $U_1, \ldots, U_s$ of U by open subsets, and for each i, some pair P_i, Q_i of polynomials such that $f = P_i/Q_i$ on U_i. In particular f is definable.*
(ii) If V is an affine variety then it turns out that the ring of regular functions on V is precisely $K[V]$. (See Proposition 4.7 in [Po 87]). Thus there is no conflict between Definition 3.4 (iii) and Definition 3.2.
(iii) We have the obvious notions of morphisms being defined over k, and the relationship to the model-theoretic notion is as in 2.7.

Definition 3.6 *Let U be a quasi-affine variety. A rational function from U to K is by definition a regular function from some nonempty open subset of U to K. If U is irreducible, one sometimes identifies two rational functions if they agree on the intersection of their domains.*

Remark 3.7 *Let V be an irreducible affine variety and U a (nonempty) open subset. Note that $RM(U) = RM(V)$. (So U is an irreducible quasi-affine variety.) By $K[U]$ we mean the ring of regular functions on U. This is easily seen to be an integral domain; $K(U)$ denotes the quotient field of $K[U]$. Equivalently $K(U)$ is the field of rational functions on U under the identification made above. Then $K(U) = K(V)$: We have a canonical homomorphism h from $K[V]$ into $K[U]$. This is an embedding, for if $P, Q \in K[X_1, \ldots, X_n]$, and $P = Q$ on U, then $P = Q$ on the Zariski closure of U which is V. Thus we have an induced embedding h of $K(V)$ in $K(U)$. But every $f \in K(U)$ has a representative of the form P/Q on some open $U_1 \subseteq U$. So $f = h(P/Q)$. If V is an affine variety, by a principal open subset of V we mean a subset defined by a single polynomial inequation $P(X) \neq 0$. Any open subset of V is then a finite union of principal open subsets.*

Remark 3.8 *Let U be a principal open subset of the affine variety V. Then, as a quasi-affine variety, U is isomorphic to some affine variety.*

Proof: Let U be defined by $P(X) \neq 0$. Let W be the affine variety $\{(x, y) : x \in V, P(x) \cdot y = 1\}$. The map $a \to (a, P(a)^{-1})$ is then clearly an isomorphism of U with W. □

Definition 3.9 *By an (abstract) variety we mean a set V with a covering by subsets $V_1, \ldots, V_m$, and for each i, a bijection $f_i : V_i \to U_i$ where U_i is an affine variety, such that for each $1 \leq i, j \leq m$*
(i) $U_{ij} = f_i(V_i \cap V_j)$ is an open subset of U_i, and
(ii) $f_j \cdot f_i^{-1}$ is an isomorphism between the quasi-affine varieties U_{ij} and U_{ji}.

Such a variety (V, V_i, f_i) is equipped with its own Zariski topology: $U \subseteq V$ is open if for each i, $f_i(U \cap V_i)$ is open in U_i. This Zariski topology is clearly *seen to be Noetherian. Again we define V to be irreducible if it is not the union of two proper closed subsets. If $(V, V_i, f_i), (W, W_j, g_j)$ are varieties then a map $f : V \to W$ is said to be a morphism if f is continuous and for any i, j, $g_j \cdot f|(f^{-1}(W_j) \cap V_i)$ is a morphism (from the quasi-affine variety $f^{-1}(W_j) \cap V_i$ onto $g_j(W_j)$).*

Remark 3.10 *(i) Any affine variety U is a variety (taking the covering of U by itself).*
(ii) By Remark 3.8, there is nothing lost if in Definition 3.9, we require the U_i to be quasi-affine varieties.
(iii) A variety (V, V_i, f_i) can, and will, be identified naturally with an object definable in K. This object is just the disjoint union of the U_i quotiented by the definable equivalence relation E, which identifies, for any ij, U_{ij} with U_{ji} via the definable map $f_j \cdot f_i^{-1}$. On the face of it this object lives in K^{eq}, but by elimination of imaginaries we can consider it as a definable subset of K^m for some m. Of course in doing this we lose the "geometry". Similarly a morphism between varieties can be identified with a definable map between the corresponding definable sets.
(iv) (V, V_i, f_i) is said to be defined over k, if the U_i and the transition maps $f_j \cdot f_i^{-1}$ are all defined over k.
(v) There is unfortunately something nonintrinsic about these definitions, as they depend on the fixed covering V_i, f_i.
(vi) We will (hopefully without ambiguity) call the variety (V, V_i, f_i) (quasi)-affine if it isomorphic to a (quasi)-affine variety.
(vii) An open or closed subset of an abstract variety has a canonical structure of a variety.
(viii) (By the product of two varieties (V, V_i, f_i) and (V, W_j, g_j), we mean $(V \times W, V_i \times W_j, (f_i, g_j))$.
(viiii) The variety V is said to be separated if the diagonal is closed in $V \times V$. Sometimes the word "variety" is reserved for separated varieties.

Definition 3.11 *(i) Suppose (V, V_i, f_i) is an irreducible variety. By a rational function $f : V \to K$ we mean a morphism (regular function) from some open subset U of V to K. Identifying rational functions if they agree on their common domain we see that the set of rational functions on V forms a field $K(V)$, the function field of V. As in Remark 3.7, if U is an affine open subset of V, then $K(V) = K(U)$.*
(ii) Irreducible varieties V, W are said to be birationally isomorphic if there is an isomorphism between open subsets of V and W.

Remark 3.12 *Suppose (V, V_i, f_i) is an irreducible variety. View V as a definable set (as above) (i) $RM(V) = tr.degree(K(V)/K)$, and $mult(V) = 1$.*
(ii) Any definable subset of V is a finite union of locally closed sets (where recall that a locally closed set is the intersection of a closed set with an open set).
(iii) If U is an open subset of V then $RM(U) = RM(V)$.

From now on we assume that all varieties we deal with are separated.

Example 3.13 *PROJECTIVE SPACE. Recall that projective n-space P^n over K is the set of lines through 0 in $K^{n+1}(= A^{n+1})$. Namely for nonzero $(x_0, \dots, x_n)$, $(y_0, \dots y_n)$, define $\bar{x} \approx \bar{y}$ if for some $\lambda \in K, \lambda\bar{x} = \bar{y}$. Then $P^n = \{\bar{x}/\approx: \bar{x} \in K^{n+1}\}$. We write a point of P^n as $(x_0 : x_1 : \dots : x_n)$ (with the obvious meaning). For each $i \leq n$, let A_i^n denote $\{(x_0 : \dots : x_n) \in P^n : x_i \neq 0\}$. Then P^n is covered by the A_i^n. For each i, let $f_i : A_i^n \to A^n$ be the bijection:*

$$f_i(x_0 : \dots : x_n) = (x_0/x_i, \dots, x_{i-1}/x_i, x_{i+1}/x_i \dots, x_n/x_i).$$

It can be checked that (P^n, A_i^n, f_i) is then a variety, defined over the prime subfield. Note that P^n is a set interpretable (over $\emptyset$) in K^{eq}, and that this set is in definable bijection (over $\emptyset$) with the $(\emptyset)$-definable set corresponding to the variety. Note also that P^n is birationally isomorphic to A^n.

A closed subset of P^n (viewed as a variety) is called a "projective variety". In fact closed subsets of P^n can also be characterised as those subsets which are zero sets of (finite) families of homogeneous polynomials. An open subset of a projective variety also has the structure of a variety. Such varieties are called quasi-projective. In fact any quasi-affine variety is quasi-projective. For affine varieties this follows from:

Remark 3.14 *Let $X \subseteq A^n$ be an affine variety. Let $Y = f_i^{-1}(X) \subseteq P^n$. Let Z be the Zariski closure of Y in P^n. Then $Y = Z \cap A_i^n$ (whereby Y is an open subset of the projective variety Z).*

Definition 3.15 *A variety X is said to be complete if for any variety Y the projection map $\pi : X \times Y \to Y$ is a closed map (in the Zariski topology), namely takes closed sets to closed sets.*

Completeness is the analogue, in the category of varieties, of the notion of compactness in the category of Hausdorff topological spaces.

Lemma 3.16 *Let X be a complete irreducible variety. Then any regular function $f : X \to K$ is constant.*

Proof: Let f be a regular function. Let $Z = \{(x, y) \in X \times K : f(x) \cdot y = 1\}$. Then Z is closed in $X \times K$, so by the completeness of X the projection $\pi(Z)$ of Z on K is closed in K. But the only closed subsets of K are K itself and the finite subsets of K. As $0 \notin \pi(Z)$, we must have that $\pi(Z)$ is finite, and thus a singleton (as $\pi(Z)$ is irreducible). So f is constant. □

Note that it follows from the above Lemma that any morphism from a complete irreducible variety to an affine variety is constant.

An important source of complete varieties is:

Fact 3.17 *Any projective variety is complete.*

Many important facts about complete varieties follow from the following result (and I thank Marc Hindry for supplying me with an easy proof).

Lemma 3.18 (The rigidity theorem) *Suppose that X, Y, Z are irreducible varieties, X is complete, $f : X \times Y \to Z$ is a morphism, and for some $y_0 \in Y, f(X \times \{y_0\})$ is a singleton. Then for each $y \in Y, f(X \times \{y\})$ is a singleton (namely for any $(x, y) \in X \times Y$, the value of $f(x, y)$ depends only on y).*

Proof: We first show that the conclusion holds for all y in some nonempty open subset of Y. Let $z_0 \in Z$ be the common value of $f(x, y_0)$. Let U be an open subset of Z which contains z_0 and is affine. As f is continuous, $V = \{(x, y) \in X \times Y : f(x, y) \notin U\}$ is closed. Thus, as X is complete, the projection V_1 of V on Y is also closed. Note that $y_0 \notin V_1$, whereby $O = Y - V_1$ is open and nonempty. For any $y_1 \in O$, the map $x \to f(x, y_1)$ is a morphism from X into U. As U is affine, it follows from 3.16 that this map is constant, as required.

Let x_0 and x_1 be in X. Then $\{y : f(x_0, y) = f(x_1, y)\}$ is a closed subset of Y, which by the previous paragraph contains the nonempty open subset O of Y, and is thus all of Y. This completes the proof. □

Finally in this section we recall the notion of smooth (or nonsingular) points on varieties. If $F \in K[X_1, \ldots, X_n]$ and $\bar{a} \in K^n$, then the differential polynomial, $d_a F$, of F at $\bar{a}$ is the linear map

$$(\delta F/\delta X_1)(\bar{a})(x_1 - a_1) + \ldots + (\delta F/\delta X_n)(\bar{a})(x_n - a_n).$$

Let $V \subseteq K^n$ be an affine variety, and suppose that $I(V)$ is generated (as an ideal) by the polynomials $F_1(\bar{X}), \ldots, F_m(\bar{X})$. Let $\bar{a} = (a_1, \ldots, a_n)$ be a point in V. The (Zariski) tangent space of V at $\bar{a}$ is defined to be

$$\{\bar{x} \in K^n : d_{\bar{a}} F_1(x) = d_{\bar{a}} F_2(\bar{x}) = \ldots = d_{\bar{a}} F_m(\bar{x}) = 0\}.$$

This is a linear subspace of K^n whose dimension is clearly seen to be $n -$ rank $(J(\bar{a}))$, where J is the $m \times n$ matrix whose i, jth entry is $(\delta F_i/\delta X_j)(\bar{a})$. With this notation we have:

Fact 3.19 *Assume V is affine irreducible with $dim(V) = m$. Then for all $\bar{a} \in V$, $rank(J(\bar{a})) \leq n - m$. Moreover $S = \{\bar{a} \in V : rank(J(\bar{a})) < n - m\}$ is a proper closed subset of V.*

Note that it is clear that for any $r, \{\bar{a} \in K^n :$ rank$(J(\bar{a})) < r\}$ is a closed set, being defined by the vanishing of all $r \times r$ minors of the matrix $J(a)$. The point is that for $r = n - m$, this closed set does not contain V. So Fact 3.19 says that for all $\bar{a}$ in some nonempty open subset of V, the dimension of the tangent space to V at $\bar{a}$ is the same as $dim(V)$. These points are called the nonsingular or smooth points of V. The set S from Fact 3.19 is called the singular locus of V. We have described smooth points in affine varieties. If V is an arbitrary irreducible variety and $a \in V$, then a is said to be a smooth point of V if a is a smooth point on some open affine subset U of V. This does not depend on the open affine set chosen. In fact smoothness of the variety V at a is a "local" issue, depending on the local ring of V at a (the ring of functions regular at a). In any case an arbitrary irreducible variety will be smooth at a nonempty open set of points. Also if f is an isomorphism of varieties V and W then f takes smooth points to smooth points. The irreducible variety V is said to be smooth or nonsingular if it is so at all its points. Similar remarks hold for reducible varieties. If V is such, then V is said to be smooth at $a \in V$, if the dimension of the tangent space to V at a is equal to max $\{dim(W) : W$ is an irreducible component of V containing $a\}$. Again V will be smooth off some closed subset of dimension strictly smaller than that of V.

4 Algebraic groups

Definition 4.1 *An algebraic group G is by definition a variety V together with a pair of morphisms $\mu : V \times V \to V$ and $\rho : V \to V$, such that μ yields a group operation on V, ρ is the map $x \to x^{-1}$. G is said to be defined over k, if V, μ, ρ are all defined over k. We will identify notationally G with V.*

From section 3 we see that an algebraic group defined over k is in a natural way a group definable in K (over k). We will prove the converse below.

Lemma 4.2 *The underlying variety of an algebraic group is separated and smooth.*

Proof: The map $f : G \times G \to G$ defined by $f(x, y) = x.y^{-1}$ is a morphism, and the diagonal is the preimage of $e \in G$ under f. For smoothness: By the remarks at the end of section 3, G is smooth at some point $a \in G$. But for any $b \in G$ there is an isomorphism (as a variety) of G taking a to b (left translation by $b.a^{-1}$), so G is also smooth at b. □

We begin with some semi-trivialities.

Lemma 4.3 *Let G be an algebraic group.*
(i) If H is a subgroup of G then the Zariski closure of H in G is also a subgroup of G.
(ii) If H is a definable subgroup of G then H is closed in G.

Proof: (i) Let X be the Zariski closure of H in G. Left multiplication by any $a \in H$ is an isomorphism of the underlying variety of G, which fixes H setwise. Thus for $a \in H, a.X = X$. Now $\{a \in X : a.X = X\}$ is (by DCC) closed in G, so contains X. Thus $a.X = X$ for all $a \in X$. So X is a subgroup of G.
(ii) Let H_1 be the Zariski closure of H. Then H_1 is a closed subgroup of G (by(i)). Now H is a finite union of sets of the form $Y - Z$ where Y is closed irreducible (and contained in H_1), and Z is a proper closed subset of Y. Let U be the union of such sets where $RM(Y)$ is maximum. Then clearly $RM(H_1 - U) < RM(H_1)$, and $U \subseteq H$. Thus H contains every generic type of H_1. The basic theory of ω-stable groups implies that $H.H = H_1$, so $H = H_1$. □

Lemma 4.4 *Let G be an algebraic group. Then the irreducible component of G which contains the identity e coincides with the smallest closed (or definable) subgroup of G of finite index. We call this subgroup G^0, the connected component of G. The other irreducible components of G are just the cosets of G^0 in G. Note that G^0 is a normal subgroup of G.*

Proof : If it is convenient to use the general theory of ω-stable groups (of which G is one) as outlined in Lascar's paper in [Las] this volume. Let G^0 be the connected component of G in the model-theoretic sense (namely the smallest definable subgroup of G of finite index). By 4.3, G^0 is closed in G. Now G^0 has multiplicity (Morley degree) 1 by [Las, 4.3]. On the other hand, by "homogeneity" of G^0, all irreducible components of G^0 have the same dimension (Morley rank), so clearly G^0 must be irreducible. The translates of G^0 in G are also irreducible. As G^0 has finite index in G, these translates together with G^0 must comprise the irreducible components of G. □

Example 4.5 *Let $GL_n(K)$ be the group of $n \times n$ invertible matrices over K. This is a principal open subset of affine n^2-space, so by Remark 3.8 is isomorphic to a closed subset of affine n^2+1 space. The group operation and inversion are clearly seen to be morphisms. Thus we can consider $GL_n(K)$ as an algebraic group whose underlying variety is affine. $GL_n(K)$ is irreducible. An algebraic group is said to be linear if it is isomorphic (as an algebraic group) to a closed subgroup of some $GL_n(k)$. It is a fact (see Lemma 4.11 of [Po 87] for a more general result) that any affine algebraic group is linear.*

Example 4.6 *An abelian variety is by definition a connected algebraic group whose underlying variety is complete.*
The group law on an abelian variety G is commutative: Consider the morphism $f : G \times G \to G$ defined by $f(g_1, g_2) = g_1^{-1}.g_2.g_1$. Note that $f(G, e) = \{e\}$. By 3.18, $f(G, g)$ is a singleton for all $g \in G$. But $g \in f(G, g)$, thus $f(G, g) = \{g\}$, for all $g \in G$, so G is commutative.
If f is a morphism from G to H where G is an abelian variety and H is a commutative algebraic group, then f is a translate of a homomorphism: Consider the map f_1 from $G \times G$ to H defined by: $f_1(x, y) = f(x) + f(y) - f(x + y)$. (Here we write the group operations in both G and H additively.) Then $f_1(0, y) = f_1(x, 0) = f(0)$ for all $x, y \in G$. By 3.18 (applied to both factors of $G \times G$), $f_1(x, y) = f(0)$ for all $x, y \in G$. Thus the map $h : G \to H$ defined by: $h(x) = f(x) - f(0)$ is a homomorphism.

Abelian varieties are discussed further in Hindry's paper in this volume. Abelian varieties of dimension 1 are precisely elliptic curves.

Fact 4.7 (Chevalley's Theorem [Ro]) *Let G be a connected algebraic group. Then G has a unique maximal normal linear closed subgroup N, and G/N is an abelian variety.*

NOTE. *For G an algebraic group and H a closed normal subgroup, it is clear (by elimination of imaginaries) that G/H is a definable group. The fact that it*

also has a (canonical) structure of an algebraic group, is more subtle, although it will follow from 4.12, below.

Remark 4.8 *We have discussed earlier the notion of a partial function from V to K being regular at a point $a \in V$ (where V is an affine variety, but the notion extends to arbitrary varieties). If $char(K) = p > 0$, and V is a variety we will call a (partial) function f from V to K p-regular at a point $a \in V$, if there is some function f_1 which is regular at a, and $f = F_r^{-m}.f_1$ for some $m > 0$. Similarly we obtain the notion of a p-rational function, and also a p-morphism between varieties.*

Lemma 4.9 *Let G, H be algebraic groups. Assume for simplicity G to be irreducible. Let $f : G \to H$ be a definable homomorphism. If $char(K) = 0$, then f is a morphism, and if $char(K) = p > 0$, then f is a p-morphism.*

Proof : For simplicity assume H affine. Suppose first $char(K) = 0$. Suppose G, H, f are all defined over k. Let U be an open subset of G which is affine (so identified with a closed subset of some K^n) and also defined over k. By 1.5, there is a nonempty open subset U_1 of U, defined over k, such that $f|U_1$ is given by a k-rational function $(P_1/Q_1, \ldots, P_m/Q_m)$, regular on U_1. Now $RM(U) = RM(G)$, so G is covered by a finite number of left translates $U_1, a_1.U_1, \ldots, a_r.U_1$ of U_1. Each $a_i.U_1$ is open in G. As f is a homomorphism, for any $i = 1, \ldots, m$, and $x \in a_iU_1, f(x) = f(a_i).f(a_i^{-1}x)$. So $f|a_i.U$ is regular. Thus f is a morphism.

If $char(K) = p$, then we obtain U_1 as above, except that now $f|U_1$ is of the form $(F_r^{-s(1)}(P_1/Q_1), \ldots, F_r^{-s(m)}(P_m/Q_m))$. The same argument as above shows that on each $a_i.U_1, f$ has a similar form, so f is a p-morphism. □

Proposition 4.10 *Any connected centerless algebraic group is linear.*

Proof : Identify some open subset of G containing e with a Zariski closed set $V \subseteq K^n$ say, such that e is the origin. Let A denote the ring of (K-valued) functions on V which are regular at e (where we identify two functions if they agree on an open set containing e). A is called the local ring of G at e. A has a unique maximal ideal $M = \{f \in A : f(e) = 0\}$. Also $\cap_n M^n = 0$. For each n, any $f \in A$ is congruent mod M^n to a polynomial of total degree at most n, so in particular A/M^n is a finite-dimensional vector space over K. Now for $g \in G$, the function $Int(g) : x \to g.x.g^{-1}$, is an automorphism of G, so, if $f \in A$, also $f^g = Int(g).f$ is in A. If $f \in M^n, f^g \in M^n$, and it is not difficult to see that actually G acts definably and K-linearly on A/M^n for all n, whereby we have a definable homomorphism $h_n : G \to GL(A/M^n)$. Now if $g \neq k \in G$, then as G is centerless, for any coordinate function π, for any open set U containing e, there is $x \in U$such that $\pi(g.x.g^{-1}) \neq \pi(k.x.k^{-1})$. So for arbitrarily large $n, \pi^g \neq \pi^h$ mod M^n. The h_n thus induce an embedding h of G in the inverse limit of the $GL(A/M^n)$. But $G_n = ker(h_n)$ is an algebraic subgroup of G. So by the DCC on such subgroups, there must be some n such that already h_n is an embedding. So we have a definable embedding of G into some $GL(n, K)$. Now use Lemma 4.9. □

We will now prove the important model-theoretic version of Weil's theorem. We may as well first state Weil's theorem as we have the language to do it. Weil's theorem (as well as the notion of an abstract variety) was originally intended and used for the construction of the Jacobian variety of a curve.

Theorem 4.11 ([We 55]) *Let V be an irreducible variety defined over the fields k. Let f be a rational function from $V \times V$ to V which is defined over k and satisfies the following:*
(i) If x, y are points of V which are generic and independent over k, and $z = f(x, y)$ then, $k(x, y) = k(y, z) = k(x, z)$ (so in particular z is also generic in V over k and is independent from each of x, y, over k).
(ii) If x, y, z are points on V which are independent and generic over k, then $f(f(x, y), z) = f(x, f(y, z))$.
Then there is a connected algebraic group G defined over k, and a birational isomorphism h from V to G defined over k, such that, for x, y in V, generic and independent over k, $h(f(x, y)) = h(x).h(y)$.

There are actually two model-theoretic versions of Weil's theorem. The first, due to Hrushovski, takes place totally within the model-theoretic framework, and states that if p is a complete stationary type (over A) in an ω-stable structure and we have an A-definable function f from $p^{(2)}$ to p, satisfying requirements much like (i) and (ii) above, then there is some A-definable function h mapping p onto the generic type of some A-definable group G, such that h takes f (at least generically) to multiplication in G. We do not prove this here. The second (due to Hrushovski and van den Dries), says that a group definable in an algebraically closed field K is definably isomorphic to an algebraic group. This is really a consequence of Theorem 4.11 (modulo a "trick" of Serre's in the positive characteristic case). In any case the proof we give below subsumes the main part of the proof of Weil's theorem, and gives additional information on "fields of definition" which is usually not mentioned by the model-theorists.

Proposition 4.12 *Let G be a group definable in the algebraically closed field K. Then G is definably isomorphic to an algebraic group. In at least the characteristic zero case, if $k < K$ and G is k-definable, then G will be k-definably isomorphic to an algebraic group which is defined over k.*

Proof: The proof will give some more information in the positive characteristic case too. We may assume G to be connected (as an ω-stable group). So both G and $G \times G$ have unique generic types over k. Let a, b be generic independent points of G over k. By 1.4,

$$a.b \in (k(a, b))_{ins} \text{ and } a^{-1} \in k(a)_{ins}.$$

We assume for now that actually

$$(*)\ a.b \in k(a.b),\ a^{-1} \in k(a) \text{ and } k \text{ is perfect.}$$

This is of course true in the characteristic 0 case, and we will show below (in Step IV) that in the positive characteristic case we can reduce to the case where (*) hold, although this may involve changing k.

STEP I. We construct a "group chunk" defined over k. Now G is a k-definable subset of K^n for some n. By quantifier elimination we can find a locally closed k-definable subset V_0 of K^n such that $V_0 \subseteq G$ and $RM(V_0) = RM(G)$. Thus $a, b, a.b$, and a^{-1} are all in V_0. Let $a.b = f(a,b), a^{-1} = h(a)$ where f, h are k-definable rational functions. By quantifier-elimination again, we can find an open subset W_0 of $V_0 \times V_0$ such that f is defined on W_0 and $f(W_0) \subseteq V_0$ (and $f(x,y) = x.y$ for $(x,y) \in W_0$).

Note that if $x \in V_0$ is generic over k, then for generic $y \in G$ over $k \cup \{x\}$, both (y,x) and $(y^{-1}, y.x)$ are generic points of $G \times G$ over k, so are in W_0. It follows from quantifier elimination that there is an open k-definable subset V_1 of V_0 such that for all $x \in V_1$, for generic $y \in G$ over $k \cup \{x\}, (y,x) \in W_0$, and $(y^{-1}, y.x) \in W_0$. Let $V = V_1 \cap V_1^{-1}$, which is a nonempty open k-definable subset of V_1. Note $V^{-1} = V$. Let $W = \{(x,y) \in W_0 : x \in V, y \in V$, and $x.y \in V\}$. W is then an open k-definable subset of $V \times V$ with the following "group chunk" features:

(**) for any $a \in V$, for generic $x \in G$ over $k \cup \{a\}$, $(x,a) \in W$ and $(x^{-1}, x.a) \in W$.

Note also that

(***) the map $(x,y) \to x.y$ from W into V is a morphism.

STEP II. Endow G with the structure of an algebraic group over $acl(k)$. Now V is a variety. As $RM(G) = RM(V)$, we can cover G by finitely many translates $V \cup a_1.V \cup \ldots \cup a_m.V$, and we try to give G the structure of a variety by means of the bijections $id : V \to V$, $a_1^{-1} : a_1.V \to V, \ldots, a_m^{-1} : a_m.V \to V$

The main technical fact is:

CLAIM. If $a', b' \in G$, then $U_{a',b'} = \{(x,y) \in V \times V : a'.x.b'.y \in V\}$ is open in $V \times V$, and the map $(x,y) \to a'.x.b'.y$ is a morphism from $U_{a',b'}$ into V.

Proof of claim. To show the claim we pick $(x_0, y_0) \in U_{a',b'}$ and show that some open neighbourhood X_0 of (x_0, y_0) is contained in $U_{a',b'}$ and that the above map restricted to X_0 is a morphism.

Let $b' = b_1 \cdot b_2$ for $b_1, b_2 \in V$. Let $c \in V$ be generic over $k \cup \{b', b_1, b_2, a', x_0, y_0\}$. Then $(c.a', x_0) \in W$ (as $c.a'$ is generic over $x_0 \in V$), and similarly $(c.a'.x_0, b_1)$, $(c.a'.x_0.b_1, b_2)$, $(c.a'.x_0.b_1.b_2, y_0)$ are in W as well as $(c^{-1}, c.a'.x_0.b_1.b_2.y_0)$. The set X_0 of pairs $(x,y) \in V \times V$ satisfying all these properties of (x_0, y_0) is seen to be open, and the map $(x,y) \to a'.x.b'.y$ to be a morphism on X_0.

From the claim we obtain that for any $c \in G, V \cap c^{-1}.V$ is open in V, and the map $x \to c.x$ from $V \cap c^{-1}.V$ to V is a morphism. Thus for each $i.j, a_i^{-1}.(a_i.V \cap a_j.V)$, and $a_j^{-1}(a_i.V \cap a_j.V)$ are both open in V and the transition map is an isomorphism. This shows that the covering by the translates of V does give G the structure of a variety. The claim says that multiplication is a morphism (and inversion is similar). This completes Step II.

STEP III. Give G the structure of an algebraic group defined over k. This is the part of Weil's theorem that is not often mentioned by model-theorists. Weil's original proof in [We 55] involved the introduction of "Chow coordinates" (i.e. eliminating imaginaries as in Lemma 1.7), normalization of varieties, and various other algebraic-geometric manipulations (including use of the Lang-Weil estimates in the case where k is a finite field!). In [We 56] an improved proof was found, based on general criteria ("descent theorem") for when a variety defined over a finite Galois extension k_1 of a field k is isomorphic to a variety defined over k. We will present without proof the descent theorem (a short proof can be found in Chapter V of [Se]).

DESCENT THEOREM Let k_1 be a finite Galois extension of a field k. Let V be a variety defined over k_1, which is either quasi-projective, or is a homogeneous space for an algebraic group defined over k_1. For $\sigma \in Gal(k_1/k)$, let V^σ be the image of V under σ. Suppose we have a system $\{h_\sigma : \sigma \in Gal(k_1/k)\}$ where h_σ is a k_1-isomorphism of V with V^σ and for $\sigma, \tau, h_{\sigma\tau} = (h_\sigma)^\tau.h_\tau$.

Then there is a variety W defined over k, and a k_1-isomorphism f of W with V such that $h_\sigma = f^\sigma.f^{-1}$ for each $\sigma \in Gal(k_1/k)$.

REMARK. It should be clear what is meant by V^σ. Namely if $V = (V, V_i, f_i)$, then by V^σ we mean $(V, V_i, \sigma.f_i)$. That is to say the U_i are replaced by U_i^σ, which are clearly also affine varieties defined over k_1.

Let us see now how the Descent Theorem yields a k-definable isomorphism of G with an algebraic group defined over k. Step II gave us the structure of an algebraic group defined over $acl(k)$ on G.

Let us denote this algebraic group by H. Let V now denote the group chunk obtained in Step II. So V "is" an open subset of H. Let k_1 be a finite Galois extension of k over which H is defined. (Remember k is perfect.) So for each $\sigma \in Gal(k_1/k), V$ is also an open subset of H^σ, so clearly the identity map on V extends to a k_1-isomorphism (which is also a group isomorphism) $h_\sigma : H \to H^\sigma$. It is clear that the h_σ satisfy the assumptions of the descent theorem. Thus we obtain an algebraic variety H_1 defined over k, and a k_1-isomorphism $f : H_1 \to H$ such that

(i) $h_\sigma = f^\sigma \cdot f^{-1}$ for all σ.

This isomorphism equips H_1 with a rational group law, which is easily seen to be defined over k. Thus H_1 is an algebraic group defined over k. Let f' be the k_1-definable isomorphism of G with H (coming from the identity map on V). It is clear that we also have

(ii) $h_\sigma = f^\sigma \cdot f^{-1}$ for all σ.

Let φ be the map $f'^{-1}.f$. Then φ is a definable isomorphism between G and H_1. On the other hand from (i) and (ii) we see that $\varphi^\sigma = \varphi$ for all $\sigma \in Gal(k_1/k)$, and so φ is k-definable. This completes Step III.

STEP IV. reduction to (*) in the positive characteristic case.

Here we destroy the question of field of definition. Assume k to be algebraically closed. We show G is k-definably isomorphic to a k-definable group for which (*) holds.

We may clearly assume k to be saturated. We know that for some $r > 0, a.b \in k(Fr^{-r}a, Fr^{-r}(b))$, and $a^{-1} \in k(Fr^{-r}(a))$. Let a be generic in G over k. Let $k^*(a)$ denote $k(Fr^{-r}(a), Fr^{-r}(a^{-1}), b.a.c, b.a^{-1}.c : b, c \in G \cap k)$. Now $k^*(a)$ is a subfield of a finitely generated extension of k, so is itself a finitely generated extension of k. So we can write $k^*(a) = k(a, c_1, \ldots, c_s)$ where $c_1, \ldots, c_s \in k^*(a)$.

Note that $c_1, \ldots, c_s \in dcl(k \cup \{a\})$. Hence we can find some k-definable $1-1$ function f from G into K^{n+s} such that $f(a) = (a, c_1, \ldots, c_s)$. Let H be the image of G under f. H is a k-definable group, definately isomorphic (over k) to G. We will show that (*) holds for H. Note that $(a, c_1, \ldots, c_s)$ realizes the generic type of H (over k). Also exchanging a with a^{-1} yields a k-automorphism of $k^*(a)$. Thus $(a, c_1, \ldots, c_s)^{-1} \in k(a, c_1, \ldots, c_s)$. Thus on the generic type of H, inversion is k-rational. We must show the same thing for multiplication. Let a_1 denote $(a, c_1, \ldots, c_s)$.

Let us remark

(i) if $b_1 \in H \cap k$, then $k(b_1, a_1) = k(a_1) = k(a_1.b_1) = (k^*(a))$.

Now let k_0 be a small algebraically closed subfield of k over which H is definable. Let $b_1 \in H \cap k$ be a k_0-generic point of H.

Clearly we have

(ii) $a_1.b_1 \in k_0(Fr^{-t}(a_1), Fr^{-t}(b_1))$ for some $t > 0$. Thus by (i), $a_1 \cdot b_1 \in k(a_1) \cap k_0(Fr^{-t}(a_1)), Fr^{-t}(b_1))$. But the latter is precisely $k_0(a_1, Fr^{-t}(b_1))$. Thus

(iii) $a_1.b_1 \in k_0(a_1, Fr^{-t}(b_1))$.

By symmetry (as $tp(a_1, b_1/k_0) = tp(b_1, a_1/k_0)$).

(iv) $a_1.b_1 \in k_0(Fr^{-t}(a_1), b_1)$.

We now use:

if a_1, b_1 are algebraically independent over the algebraically closed field k_0 then $k_0(Fr^{-t}(a_1), b_1) \cap k_0(a_1, Fr^{-t}(b_1)) = k_0(a_1, b_1)$.

Thus $a_1.b_1 \in k_0(a_1, b_1)$.

We have (*) as required. □

Finally:

Proposition 4.13 *Let F be an infinite field definable in K. Then F is (in $(K, +, ., 0, 1)$) definably isomorphic to K.*

Proof: We will use a number of basic facts on linear algebraic groups, which can be found in Borel [Bor] or Springer [Sp]. Note that $(F, +, .)$ is ω-stable, so by the result in the Appendix, it is algebraically closed, and moreover both the additive group and multiplicative groups of F are connected. Let G be the semidirect product of $(F, +)$ with $(F, .)$ (where $(F, .)$ acts on $(F, +)$ by multiplication in the field F). Then G is a centerless group definable in K. Let G^+ be the definable subgroup $(F, +)$. G^+ is connected. By 4.11 and 4.9, we may assume that G is an algebraic subgroup of $GL(n, K)$ for some n. Then G^+ is a connected commutative algebraic linear group. The Jordan decomposition yields that G^+ is the direct product of a unipotent group and an algebraic torus. An algebraic torus is divisible and has elements of arbitrarily large finite exponent. As G^+ is abstractly isomorphic to the additive group of an algebraically closed field, this forces G^+ to be unipotent. In characteristic $p > 0$, a commutative unipotent group has exponent some power of p, while in characteristic 0, a commutative unipotent group is isomorphic (as an algebraic group) to a vector space over K. This forces $char(F) = char(K)$, and G^+ to have exponent p in the $char(K) = p$ case. Finally, a commutative unipotent group of characteristic

p is again isomorphic (as an algebraic group) to a vector space over K. Thus, we can in all characteristics, identify G^+ with a (finite-dimensional) vector space over K. We have shown that we can identify definably $(F, +)$ with $(K^n, +)$ for some n. □

The following is immediate in characteristic 0, and a little less so in positive characteristic:
FACT. Any definable group of automorphisms of K^n is linear.

Let e be the multiplicative unit of F and o the multiplication on F. Denote by . scalar multiplication of K on K^n. We define a map h from K into $F = K^n$: $h(\lambda) = \lambda.e$. We remark that h is 1−1, and that $(\lambda.e)o(\mu.e) = \lambda\mu(eoe) = \lambda\mu.e$, by linearity. Thus h is a field embedding. It follows that F is definably isomorphic to a finite extension of K, and thus to K itself, as K is algebraically closed.

5 ω-stable fields.

This section is only concerned with ω-stable groups (which of course include ω-stable fields), is independent of the first four sections, and depends on the machinery developed by Lascar in his paper [Las] in this volume. By an ω-stable field, we will mean here a structure which is an expansion of a field and whose theory is ω-stable.

Lemma 5.1 *Suppose F is an infinite ω-stable field. Then F has unique generic type.*

Proof: Note that, by definition, a generic type of F is a type $p(x) \in S_1(F)$ such that $RM(p) = RM(F)$. On the other hand, by Proposition 4.3 of [Las], F has a unique generic type if F is connected (additively or multiplicatively). Suppose by way of contradiction that F had a proper additive subgroup A of finite index. By the ω-stable DCC, $I = \{a.A : a \in F\}$ is equal to a finite subintersection $a_1.A \cap \ldots \cap a_n.A$. But the latter clearly also has finite index in F, and is an ideal, so equals F, a contradiction. □

We will be assuming the following basic fact about $acl(-)$ in ω-stable groups.
FACT. Suppose G is an ω-stable group, $A \subseteq G$ and $a_1, \ldots, a_n$ are generic independent elements of G over A. Suppose that $(a_1, \ldots, a_n) \in acl(b_1, \ldots, b_n, A)$ where the b_i are also in G. Then $b_1, \ldots, b_n$ are also generic independent over A.

Proposition 5.2 *Any infinite ω-stable field is an algebraically closed field.*

Proof: We may assume F to be ω-saturated.
CLAIM (i). F is perfect.
Proof of the Claim: Assume $char(F) = p > 0$(otherwise there is nothing to do). $Fr : F \to F$ is an endomorphism of $(F, +)$, so the Fact above implies that the image of Fr has the same Morley rank as F. Thus F^p is a subgroup of F of finite index, so equals F by connectedness.

Let us fix some algebraic closure K of F.

CLAIM (ii). Let $a_0, \ldots, a_{n-1}$ be generic independent (over $\emptyset$) in F. Then all solutions of
(*) $X^n + a_{n-1}X^{n-1} + \ldots + a_1X + a_0$
in K are already in F.
Proof of the Claim: Let $b_0, \ldots, b_{n-1}$ be generic independent in F over $\emptyset$. Let $c_0, \ldots, c_{n-1}$ be the symmetric functions in $b_0, \ldots, b_{n-1}$, then the b_i are the roots of $X^n + c_{n-1}X^{n-1} + \ldots + c_1X + c_0$. The $c_i \in F$, and $b_0, \ldots, b_{n-1} \in acl(c_0, \ldots, c_{n-1})$ in F. Thus (by the above Fact) $c_0, \ldots, c_{n-1}$ are independent generic elements of F over $\emptyset$. Uniqueness of the generic type of F implies that $tp(a_0, \ldots a_{n-1}/\emptyset) = tp(c_0, \ldots, c_{n-1}/\emptyset)$. Thus by automorphism (*) has n distinct solutions in F, proving the claim.

Now let $P(X) = X^n + a_{n-1}X^{n-1} + \ldots + a_1X + a_0$ be the minimal polynomial of some element $\alpha \in K$. Assume for a contradiction that $\alpha \notin F$, namely $n > 1$. As F is perfect P has distinct roots $\{\alpha_1, \ldots, \alpha_n\}$. So $L = F(\alpha_1, \ldots, \alpha_n)$ is a Galois extension of F, and moreover the $n \times n$ matrix whose ith row is $[1, \alpha_i, \alpha_i^2, \ldots, \alpha_i^{n-1}]$ is invertible. Let $t_0, \ldots, t_{n-1} \in F$ be generic independent over $\{a_0, \ldots, a_{n-1}\}$, and let
(**) $r_i = t_0 + t_1\alpha_i + t_2\alpha_i^2 + \; + t_{n-1}\alpha_i^{n-1}$.

Let $c_0, \ldots, c_{n-1}$ be the elementary symmetric functions in $r_1, \ldots, r_n$. Then (as each c_i is fixed by $Gal(L/F)$), the c_i are in F. Now we have
(iii) each t_i is in the field generated by $\alpha_1, \ldots, \alpha_n, r_1, \ldots, r_n$.
(iv) (working in L) each r_i is field-theoretically algebraic over $\{c_0, \ldots, c_{n-1}\}$.

It follows from (iii) and (iv) that in F, for each i
(v) $t_i \in acl(a_0, \ldots, a_{n-1}, c_0, \ldots, c_{n-1})$.

The above Fact again implies that $c_0, \ldots, c_{n-1}$ are generic, independent (over $\emptyset$) elements of F. By (ii) $r_1, \ldots, r_n$ are in F. But then (**) contradicts the fact that the minimal polynomial of α over F has degree n. Thus $\alpha \in F$. We have shown that F is algebraically closed. □

References

[Bor] A. Borel, *Linear Algebraic Groups*, 2nd. edition, Springer, 1991.

[Bous1 89] E. Bouscaren, *Model-theoretic versions of Weil's theorem on pregroups*, in The Model Theory of Groups, A. Nesin and A. Pillay ed., Notre Dame University Press 1989.

[Lan] S. Lang, Algebra, Addison-Wesley.

[Las] D. Lascar, *ω-stable groups*, this volume.

[Pi 89] A. Pillay, *Model theory, stability and stable groups*, in The Model Theory of Groups, A. Nesin and A. Pillay ed., Notre Dame University Press, 1989.

[Pi2 96] A. Pillay, *Geometrical Stability Theory*, Oxford University Press, 1996.

[Po 87] B. Poizat, Groupes Stables, Nur al-matiq wal ma'rifah, Villeurbanne, France, 1987.

[Po 89] B. Poizat, *An introduction to algebraically closed fields*, in The Model Theory of Groups, A. Nesin and A. Pillay ed., Notre Dame University Press, 1989.

[Ro] M. Rosenlicht, *Some basic theorems on algebraic groups*, American Journal of Math. 78 (1956), 401-443.

[Se] J.P. Serre, *Algebraic Groups and Class Fields*, Springer, 1988.

[Sh] I.R. Shafarevich, *Basic Algebraic Geometry*, Springer, 1977.

[Sp] T.A. Springer, *Linear Algebraic Groups*, Birkhauser, 1981.

[We 62] A. Weil, *Foundations of Algebraic Geometry*, AMS 1962.

[We 55] A. Weil, *On algebraic groups of transformations*, American Journal of Math. 77 (1955).

[We 56] A. Weil, *The field of definition of a variety*, American Journal of Math. 78 (1956), 509-524.

[Zie] M. Ziegler, *Introduction to stability theory and Morley rank*, this volume.

Introduction to abelian varieties and the Mordell-Lang conjecture

MARC HINDRY

Our aim in this brief survey is to try and give an intuition about abelian varieties to model theorists, working mainly over the complex field. We then try to motivate the Lang conjecture, also called the Mordell-Lang conjecture, showing especially how it generalizes Mordell's conjecture.

Finally we present an analogue over function fields, ending with the statement of the result proved by E. Hrushovski [Hr 96], which is explained at the end of this volume in [Bous].

For basic definitions of varieties, algebraic groups, completeness, morphisms see for example [Sh] or, for a model theoretic presentation [Pil] in this volume.

1 Abelian varieties

Definition 1.1 *An* abelian variety *is a complete algebraic group.*

Remark: By convention, we include connectedness in the definition of an abelian variety, similarly, when we talk of a variety, we mean an irreducible variety. An abelian subvariety of A is a closed connected subgroup of A.

Over the complex numbers, one can give an apparently totally different definition. Recall that a *lattice* in $\mathbb{C}^g$ is a discrete subgroup of maximal rank, hence isomorphic to $\mathbb{Z}^{2g}$.

Definition 1.2 *A* complex torus *is a quotient $\mathbb{C}^g/L$, where L is a lattice in $\mathbb{C}^g$. A* complex abelian variety *is a complex torus $\mathbb{C}^g/L$, equipped with a non degenerate* Riemann form, *that is a hermitian positive definite form $H : \mathbb{C}^g \times \mathbb{C}^g \to \mathbb{C}$ such that $ImH : L \times L \to \mathbb{Z}$.*

It is of course quite a deep theorem that the two definitions agree (when the ground field is $\mathbb{C}$). We will just show here that an abelian variety over $\mathbb{C}$ is a torus.

Example ($dimA = 1$)
Consider a smooth cubic projective curve given by the equation $ZY^2 = X^3 + aXZ^2 + bZ^3$ where $4a^3 + 27b^2 \neq 0$. We take as origin the point "at infinity" $[0, 1, 0]$ and define the group law by the tangent and chord process described in

the figure below. Of course the drawing represents the curve in affine coordinates $x = X/Z$, $y = Y/Z$.

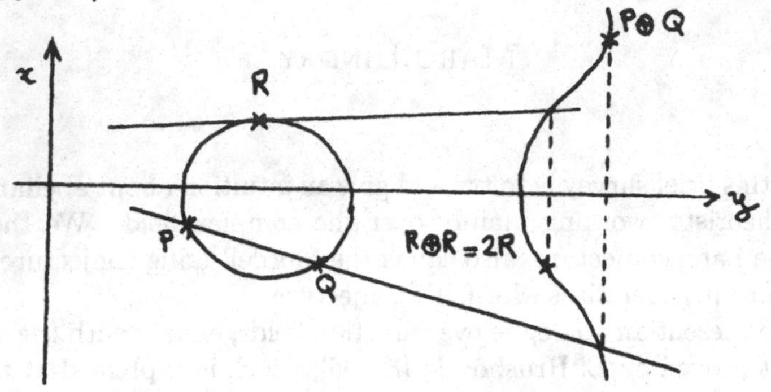

This is an abelian variety of dimension 1, an elliptic curve. We will see more examples shortly.

Lemma 1.3 *A complex abelian variety is commutative.*

Proof: In this proof, and only in this proof, we write multiplicatively the group law on A, an abelian variety. Consider the map from A to A defined by $\phi_a(x) = axa^{-1}$; let $d\phi_a$ be the differential at the origin, hence $d\phi_a$ is a linear endomorphism of the tangent space V of A at the origin. We get thus a holomorphic map from A to $End(V) \cong \mathbb{C}^{g^2}$ which maps a to $d\phi_a$. By the maximum principle or Liouville's theorem, the map is constant and its value is Id_V. It follows that $\phi_a = Id_A$ and hence A is commutative. □

For an algebraic proof, one replaces compactness by completeness (see for example [Pil, 4.6]).

Lemma 1.4 *A complex abelian variety is a torus*

Proof: One actually shows that a compact complex Lie group is a complex torus. We saw in the previous lemma that it is commutative, hence the exponential $exp_A : V \to A$ is a surjective homomorphism. Its kernel L is then a discrete subgroup and, since A is compact, L has to be of rank $2g$. □

It can also be shown that all abelian varieties are projective varieties, that is admit an embedding $A \to \mathbb{P}^n$ as a closed subvariety.

Let us now review some classical properties of abelian varieties. We give the easy proofs over $\mathbb{C}$ and just state the results over an arbitrary algebraically closed field K. The extension of the results over an algebraically closed field of characteristic zero is usually straightforward. The extension of the results over an algebraically closed field of characteristic p usually requires entirely different techniques stemming from algebraic geometry.

Let K be an arbitrary algebraically closed field and A an abelian variety over K.

1. A is a divisible group: let $[n]$ denote multiplication by n, then the map $[n] : A \to A$ is surjective.

   Indeed we have a commutative diagram

$$\begin{array}{ccc} \mathbb{C}^g & \xrightarrow{n} & \mathbb{C}^g \\ \downarrow & & \downarrow \\ \mathbb{C}^g/L & \xrightarrow{n} & \mathbb{C}^g/L \end{array}$$

   This remains true over any algebraically closed field K. The proof is essentially the "same" if $char(K)$ does not divide n and presents extra difficulties if $char(K)$ divides n (see [Mu]).

2. Torsion points are dense and are algebraic over the field of definition of A. Over $\mathbb{C}$ one gets

$$A_{\text{torsion}} = \mathbb{Q}L/L \cong (\mathbb{Q}/\mathbb{Z})^{2g} \subset (\mathbb{R}/\mathbb{Z})^{2g} \cong A(\mathbb{C})$$

   in fact more precisely $Ker[n] = \frac{1}{n}L/L \cong (\mathbb{Z}/n\mathbb{Z})^{2g}$.

   The situation over an arbitrary algebraically closed field K is as follows, for an abelian variety A of dimension g:

$$Ker[\ell^n] \cong \begin{cases} (\mathbb{Z}/\ell^n\mathbb{Z})^{2g}, & \text{if } char(K) \neq \ell \\ (\mathbb{Z}/\ell^n\mathbb{Z})^{r}, & \text{if } char(K) = \ell \quad (\text{with } 0 \leq r \leq g) \end{cases}$$

   If A is defined over $k \subset K$, then so is the morphism $[n]$ and its kernel, hence the points of $Ker[n]$ are defined over the algebraic closure of k. If n is prime to $p = char(K)$, then the points of $Ker[n]$ are defined over the separable closure of k.

   Corollary: (Strong rigidity) If A is defined over a field k, then all closed subgroups of A are defined over the separable closure of k.

   Indeed if $G \subset A$ is such a closed subgroup, then $G \cap A_{\text{torsion}}$ is dense in G and composed of points defined over the algebraic closure of k; if $char(k) = p$ we take only the torsion of order prime to p to stay in the separable closure.

3. (Poincaré's reducibility theorem) If B is an abelian subvariety of an abelian variety A, there exists another abelian subvariety C such that $A = B + C$ and $B \cap C$ is finite (in other words the map from $B \times C$ to A defined by $(b, c) \mapsto b + c$ is an isogeny).

   Proof: Let $A = V/L$ (with L a lattice in $V \cong \mathbb{C}^g$) be an abelian variety. It possesses a non degenerate Riemann form, say H. Let W be the complex tangent space of B at 0, then $B = W/L \cap W$. Let W' be the orthogonal complement of W with respect to H, then one shows that $C = W'/L \cap W'$ is a complex torus and that H induces a non degenerate Riemann form on it. Since $V = W \oplus W'$, the abelian subvariety C does what is required. □

Remarks: a) This property is false (in general) for complex tori; on the other hand it remains valid for all abelian varieties defined over a field of any characteristic.

b) One can deduce from Poincaré's reducibility theorem that quotients of A and abelian subvarieties of A are the same up to isogeny. Indeed, if $f : A \to B$ is onto, then the connected component of the kernel of f has a "complementary" abelian subvariety C such that the induced map $f : C \to B$ is an isogeny. For the converse, if B is an abelian subvariety, C a "complementary" abelian subvariety then A/C is isogenous to B.

4. The endomorphism ring $\mathrm{End}(A)$ is isomorphic to some $\mathbb{Z}^r$ with $r \leq 4g^2$ (as a group).

Proof: Any endomorphism $\alpha : A = \mathbb{C}^g/L \to A = \mathbb{C}^g/L$ induces a linear map $\tilde{\alpha} : \mathbb{C}^g \to \mathbb{C}^g$ such that $\tilde{\alpha}(L) \subset L$ hence we get an injection $\mathrm{End}(A) \hookrightarrow \mathrm{End}_{\mathbb{Z}}(L) \cong Mat(2g \times 2g, \mathbb{Z})$. □

This property remains true in any characteristic. From this last property or from the strong rigidity property, one easily deduces:

Corollary: In an abelian variety, there is no (non constant) algebraic family of abelian subvarieties.

Remark: This is somewhat trivial but quite necessary for the plausibility of Lang's conjecture.

Example: An abelian variety A is called *simple* if its only abelian subvarieties are $\{0\}$ or itself; in this case $\mathrm{End}(A) \otimes \mathbb{Q}$ is a division ring.

The fundamental example of an abelian variety : the jacobian of a curve.

Historically the theory comes from integral calculus, a typical example being attempts to compute integrals like:

$$\int \frac{dx}{\sqrt{x^3+1}}, \int \frac{dx}{\sqrt{x^5+1}}.$$

Since all algebraic curves are birationally equivalent to a smooth projective curve, we will restrict to the latter.

Definition of the genus of a curve: the complex points of a smooth projective curve X form a compact Riemann surface $X(\mathbb{C})$, hence one of the following shape

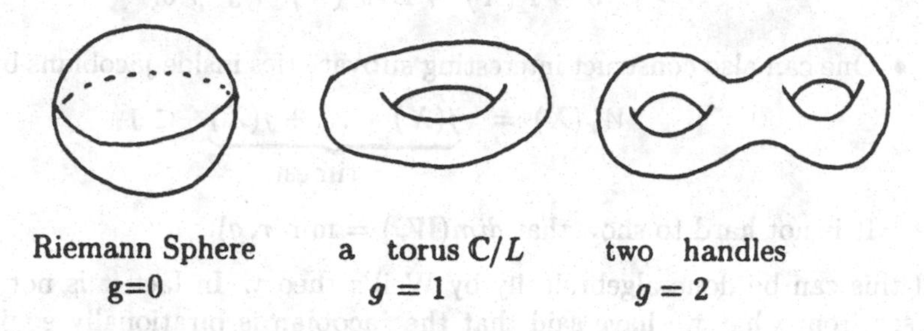

The genus can also be defined purely algebraically as the number of linearly independant *regular differential 1-forms*. Selecting a point $P_0 \in X(\mathbb{C})$ and $\omega_1, \ldots, \omega_g$ such differential forms we may define

$$P \mapsto \left(\int_{P_0}^{P} \omega_1, \ldots, \int_{P_0}^{P} \omega_g \right)$$

this is actually ill-defined because the integral will depend on the chosen path

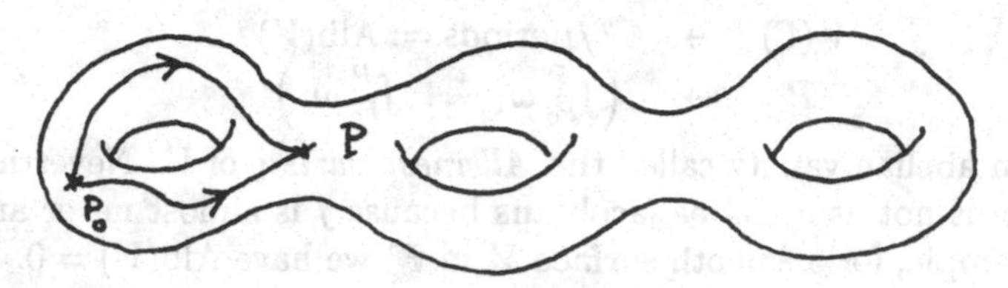

but if we mod out by the periods – integrals of the ω_i's along closed paths – we get a well-defined holomorphic map

$$j : X(\mathbb{C}) \to \mathbb{C}^g/\text{periods} =: \mathrm{J}(\mathbb{C}).$$

The torus J is called the *jacobian* of X and denoted $\mathrm{J}(X)$. A beautiful theorem of Riemann – Riemann's periods relations (see [BiLa] or [Bos]) – guarantees that J is an abelian variety, in particular the periods generate a lattice. From our point of view the main results concerning the jacobian are:

- If $g \geq 1$ then $j : X \hookrightarrow \mathrm{J}$ is an embedding; it is an isomorphism if and only if $g = 1$ (in fact $dim(\mathrm{J}) = g$).
- (Abel-Jacobi) We may extend the map j by linearity to *divisors* (i.e. formal sums of points of X). For a rational (meromorphic) function f on X, define its divisor by $div(f)$ ={Zeroes of f} - { Poles of f} (see [Sh] for a precise definition of multiplicities). The *degree* of a divisor $D = \sum_i n_i P_i$ is defined as $deg(D) := \sum_i n_i$. The divisors of functions form a subgroup $P(X)$ of the group of divisors of degree zero $Div^0(X)$. The Abel-Jacobi

theorem may be summarised by saying that the following canonical sequence is exact:

$$0 \to P(X) \to Div^0(X) \to \mathrm{J} \to 0.$$

- One can also construct interesting subvarieties inside jacobians by setting:

$$W_r(X) := \underbrace{j(X) + \ldots + j(X)}_{r \text{ times}} \subset \mathrm{J}$$

It is not hard to show that $dim(W_r) = \min(r, g)$.

All this can be done algebraically by Weil's theory. In fact it is not too hard to see from what we have said that the jacobian is birationally equivalent to the g-th symmetric product of X. The classical theorem of Riemann-Roch provides a birational group law on this symmetric product. Weil invented the theorem quoted in [Pil, 4.11] in order to get a purely algebraic construction of the jacobian. In particular his construction is valid over a field of characteristic p (see [Se]).

Remark: This construction may be generalized to an arbitrary smooth projective variety V by selecting $\omega_1, \ldots, \omega_g$ a basis of holomorphic 1-forms and setting:

$$\begin{array}{rccl} j: & V(\mathbb{C}) & \to & \mathbb{C}^g/\text{periods} =: \mathrm{Alb}(V) \\ & P & \mapsto & \left(\int_{P_0}^P \omega_1, \ldots, \int_{P_0}^P \omega_g\right) \end{array}$$

One still gets an abelian variety called the *Albanese variety* of V. Nevertheless, the construction is not as useful as jacobians because j is almost never an embedding; for example, for a smooth surface V in $\mathbb{P}^3$ we have $\mathrm{Alb}(V) = 0$.

Anticipating, we may note that Lang's conjecture says something only for varieties which admit regular differential 1-forms.

We focus in this paper on abelian varieties but it is interesting to study general (commutative) algebraic groups. We will only mention Chevalley's theorem and the definition of semi-abelian varieties.

Definition 1.5 *An affine algebraic group is an affine variety (i.e. closed in an affine space) with an algebraic group law.*

Examples:

- $GL(n)$ is an affine algebraic group: instead of $\{x \in Mat(n \times n) \mid det(x) \neq 0\}$, think of it as $\{(x,t) \in Mat(n \times n) \times \mathbb{A}^1 \mid tdet(x) - 1 = 0\}$.
- Over an algebraically closed field, the only affine algebraic groups of dimension one are: the additive group $\mathbb{G}_a$ with underlying variety the affine line $\mathbb{A}^1$ and group law the addition; the multiplicative group $\mathbb{G}_m$ with underlying variety the affine line minus a point $\mathbb{A}^1 \setminus \{0\}$ and group law the multiplication.

- Any closed subgroup of $GL(n)$ is an affine algebraic group. In fact it can be shown that any affine algebraic group is isomorphic to such a closed subgroup of $GL(n)$.

It is easy to see that the compositum of two affine subgroups of an algebraic group G is again affine, hence there is a *unique maximal connected affine subgroup.*

Theorem 1.6 (Chevalley) *Let G be a connected algebraic group, let L be the maximal connected affine subgroup of G, then G/L is an abelian variety.*

Remark: L is also the smallest closed subgroup of G such that G/L is an abelian variety.

Definition 1.7 *Over an algebraically closed field, a semi-abelian variety is a commutative algebraic group which is an extension of an abelian variety by a multiplicative group $(\mathbb{G}_m)^r$.*

One can formulate (and prove ... see papers by Faltings and Vojta) Lang's conjecture for the wider class of semi-abelian variety. Many properties of abelian varieties are shared by semi-abelian varieties. For example, if G is a semi-abelian variety defined over an algebraically closed field K:

- Torsion point are dense in G
- All closed subgroups are defined over K
- If G is an extension of A with dimension a by $(\mathbb{G}_m)^r$ then $Ker[n]_G \cong (\mathbb{Z}/n\mathbb{Z})^{2a+r}$ as long as $char(K)$ does not divide n.

2 Lang's conjecture

The conjecture of Lang stems from Mordell's conjecture (1922) and a question raised by Manin and Mumford in the sixties. Mordell's conjecture is a problem about diophantine equations. Given a polynomial $P \in \mathbb{Q}[x,y]$, one wants to study the set

$$\{(x,y) \in \mathbb{Q}^2 \mid P(x,y) = 0\}.$$

Consider the associated smooth projective curve X, which is birational to the affine curve defined by $P(x,y) = 0$.

- 1st case : The curve X is of genus 0. Then either $X(\mathbb{Q}) = \emptyset$ (e.g. $P(x,y) = x^2 + y^2 + 1$) or all but finitely many solutions are parametrised by rational fractions $x(t), y(t)$ (e.g. all the solutions of $x^2+y^2-1=0$ are parametrised by $(x,y) = (\frac{2t}{t^2+1}, \frac{t^2-1}{t^2+1})$ except $(0,1)$).
- 2nd case : The genus of X is one. Then either $X(\mathbb{Q}) = \emptyset$ or, taking one of the rational points as origin, X is an elliptic curve (an abelian variety of dimension 1) and hence we have a group law on the set $X(\mathbb{Q})$.

Theorem 2.1 (Mordell-Weil) *$X(\mathbb{Q})$ is a finitely generated group. Much more generally, if K is a field finitely generated over $\mathbb{Q}$ (for example a number field) and A is an abelian variety defined over K then $A(K)$ is a finitely generated group.*

- 3rd case : The genus of X is ≥ 2. Then Faltings proved the Mordell conjecture : $X(\mathbb{Q})$ is finite.

We can reformulate the Mordell conjecture as follows : embed $X \hookrightarrow A =$ jacobian of X, then $X(\mathbb{Q}) = X \cap A(\mathbb{Q})$, so we are reduced to proving:
Let Γ be a finitely generated subgroup of $A = \mathrm{J}(X)$ then $\Gamma \cap X$ is finite
or slightly more generally:
Let X be a curve in an abelian variety A and Γ be a finitely generated subgroup of A then $\Gamma \cap X$ is finite, except if X is a translate of an elliptic curve.
Granting the Mordell-Weil theorem, this last statement is actually easily seen to be equivalent to Mordell's conjecture.
The *Manin-Mumford question* is the following : consider again $X \hookrightarrow A = \mathrm{J}(X)$ then is it true that $X \cap A_{\mathbf{torsion}}$ is finite? Analogously one can slightly generalize this to:
Let X be a curve in an abelian variety A, then $X \cap A_{\mathbf{torsion}}$ is finite, except if X is a translate of an elliptic curve.
Remark: The motivations were apparently quite different. Manin had proven with Drinfeld that cusps generate a torsion subgroup of the jacobian of a modular curve. Mumford was studying moduli spaces of curves.

It is easy to put together the two conjectures by introducing the concept of a group of finite rank.

Definition 2.2 *Γ is a group of finite rank r if $\Gamma \otimes \mathbb{Q} \cong \mathbb{Q}^r$.*

Or, if you prefer, there is a finitely generated subgroup Γ_0 of rank r, such that for all $\gamma \in \Gamma$, there is an integer $m \geq 1$ such that $m\gamma \in \Gamma_0$.
Note : When working in characteristic p, one requires a bit more, namely that $\Gamma \otimes \mathbb{Z}_{(p)} \cong \mathbb{Z}_{(p)}^r$ or, if you prefer, there is a finitely generated subgroup Γ_0 of rank r, such that for all $\gamma \in \Gamma$, there is an integer $m \geq 1$, coprime with p, such that $m\gamma \in \Gamma_0$.

Lang's conjecture for curves *Let X be a complex curve in an abelian variety A and Γ be a subgroup of finite rank in A then $\Gamma \cap X$ is finite, except if X is a translate of an elliptic curve.*

It is natural to ask what happens for higher dimensional varieties.

Lang's conjecture (the "absolute form", characteristic zero) *Let X be a complex subvariety of a complex abelian variety A and Γ be a subgroup of finite rank in A then there exist $\gamma_1, \ldots, \gamma_m \in \Gamma$ and $B_1, \ldots, B_m$ abelian subvarieties such that $\gamma_i + B_i \subset X$ and such that*

$$\Gamma \cap X(\mathbb{C}) = \cup_{i=1}^{m} \gamma_i + (B_i(\mathbb{C}) \cap \Gamma).$$

Remarks:

- The most immediate analog over $k = \overline{\mathbb{F}}_p$ is false since all points in $A(\overline{\mathbb{F}}_p)$ are torsion points. Nevertheless, there is a relative form of Lang's conjecture, which we will state in the next section.
- This is really a diophantine conjecture. Indeed if $K = \mathbb{Q}(x_1, \ldots, x_n)$ is a field finitely generated over $\mathbb{Q}$ (e.g. a number field), if X is a subvariety of an abelian variety A, all defined over K, then the Mordell-Weil theorem tells us that $\Gamma = A(K)$ is finitely generated, hence the conjecture describes the K-rational points of X.
- The conjecture is easily equivalent to the following statement: *if the set $X \cap \Gamma$ is Zariski dense in X then X is a translate of an abelian subvariety by a point in Γ* (proof : by induction on the dimension of X).
- One may perform a number of reduction in order to prove the conjecture. For example if we define the ***stabilizer*** of X as the (not necessarily connected) algebraic subgroup $\mathrm{Stab}_X = \{a \in A \mid a + X \subset X\}$ then we may assume that Stab_X is finite or even $\{0\}$. Notice that under the hypothesis that Stab_X is finite, the conclusion must be that $X \cap \Gamma$ is not dense in X.
  Proof: Call $H = \mathrm{Stab}_X$ and consider the abelian variety $A' := A/H$ with the canonical projection $\pi : A \to A'$ and image $X' := \pi(X)$. One checks easily that $\mathrm{Stab}_{X'} = \{0\}$. Hence, if X is not a translate of an abelian subvariety, X' is not reduced to a point. If we already know that $X' \cap \pi(\Gamma)$ is not Zariski dense in X', we immediately obtain that $X \cap \Gamma$ is not Zariski dense in X, since $\pi(X \cap \Gamma) \subset X' \cap \pi(\Gamma)$. □
- Though the formulation is given over the field of complex numbers, one may assume that A, X and Γ_0 are defined over a field finitely generated over $\mathbb{Q}$ (here Γ_0 is a finitely generated subgroup such that for each $\gamma \in \Gamma$, there is an $n \geq 1$ such that $n\gamma \in \Gamma_0$). By specialisation arguments, one may reduce the conjecture to the case where everything is defined over $\overline{\mathbb{Q}}$.

To illustrate the richness of the content of Lang's conjecture, let us get back to the case of a curve X defined over (say) a number field K and embedded in its jacobian $A = \mathrm{J}(X)$ and consider the subvarieties $W_r = X + \ldots + X$ introduced earlier. The Lang conjecture asserts that $W_r(K)$ is finite unless W_r contains a translate of a non zero abelian subvariety. On the other hand there is an obvious morphism from the r-th symmetric product of the curve X with itself onto W_r. The Abel-Jacobi theorem tells us that this is an isomorphism except when X admits a morphism of degree $\leq r$ to $\mathbb{P}^1$. Since points defined over K on the symmetric product correspond essentially to algebraic points on X with Galois orbit (over K) of cardinal $\leq r$, we get the following result (see [Hi 88], [Hi 92] for details):

Proposition 2.3 *Let X be a curve defined over a number field K and let r be an integer. Assume that there is no morphism $\pi : X \to \mathbb{P}^1$ of degree $\leq r$ and also that W_r does not contain a translate of a non zero abelian subvariety. Then the union of the set of rational points $X(L)$, for all number fields L containing K with $[L : K] \leq r$, is finite.*

One may formulate obvious generalizations to finitely generated fields. The condition " no morphism $\pi : X \to \mathbb{P}^1$ of degree $\leq r$" is necessary since otherwise the set $\pi^{-1}(\mathbb{P}^1(K))$ is clearly infinite. One would like a geometric description "in terms of the curve" of the condition "W_r does not contain a translate of a non zero abelian subvariety", but that does not seem to be known in general.

Let us digress a bit to explain how the Lang conjecture (for subvarieties of abelian varieties) fits into some general conjecture about classification of algebraic varieties and diophantine properties.

Let V be a smooth projective variety of dimension n, we have already considered regular differential 1-forms, but we may also introduce r-forms (say for $1 \leq r \leq n$) and also powers of r-forms. We refer for example to Shafarevic's book [Sh] and mention only the abstract definitions: considering the *sheaf* Ω_r of differential r-forms, a *regular differential r-form of weight m* is a global section of $\Omega_r^{\otimes m} = \Omega_r \otimes \ldots \otimes \Omega_r$. The most interesting case (for our purposes) is the case of n-forms (recall $n = dim(V)$) because in this case the sheaves $\Omega_n^{\otimes m}$ are invertible and give rises to linear systems. The sheaves $\Omega_n^{\otimes m}$ are called the *pluri-canonical sheaves* and the dimensions g_m of the spaces of global sections are called the *pluri-genera* of the variety V. More concretely select a basis $\omega_1, \ldots, \omega_{g_m}$ of global sections of $\Omega_n^{\otimes m}$, then the map

$$\begin{array}{cccc} \Phi_m : & V & \cdots\to & \mathbb{P}^{g_m - 1} \\ & x & \to & (\omega_1(x), \ldots, \omega_{g_m}(x)) \end{array}$$

is a rational map and a morphism outside the set of common zeroes of the ω_i. Observe that the "value" of ω_i at x is meaningless but the quotient ω_i/ω_j is a function on V. So we have a collection of rational maps called quite naturally the *pluri-canonical maps.* This is used to define an invariant which is fundamental in the classification of algebraic varieties:

Definition 2.4 *Let V be a smooth projective variety, the Kodaira dimension of V is $\kappa(V) = -1$ if all $g_m(V) = 0$ and $\kappa(V) = \max\{dim(\Phi_m(V)\}$ otherwise. Clearly $-1 \leq \kappa(V) \leq dim V$.*

Examples: For a curve V it is easy (or at least classical) to see that $\kappa(V) = -1$ if the genus is zero ($V \cong \mathbb{P}^1$), $\kappa(V) = 0$ if the genus is 1 and $\kappa(V) = 1$ if the genus is at least 2. An abelian variety A has $\kappa(A) = 0$. A smooth hypersurface V of degree d in $\mathbb{P}^n$ has $\kappa(V) = -1$ if $d \leq n$ and $\kappa(V) = 0$ if $d = n + 1$ and $\kappa(V) = n - 1 = dim(V)$ if $d \geq n + 2$. For a subvariety V of an abelian variety A, it is known that $\kappa(V) = dim(V) - dim(\mathrm{Stab}_V)$.

Definition 2.5 *A variety is of general type if $\kappa(V) = dim(V)$.*

Thus a subvariety of an abelian variety is of general type if and only if its stabilizer is finite; a smooth hypersurface of degree d in $\mathbb{P}^n$ is of general type if and only if $d \geq n+2$.

Conjecture (Bombieri-Lang) *Let V be a variety of general type defined over a number field K then the set of rational points $V(K)$ is not Zariski dense.*

Remarks: 1) Bombieri asked this question for surfaces of general type and Lang (independantly) made the general conjecture in a more precise form: he conjectures that there is a fixed "geometric" closed subset Z such that, if $U = V \setminus Z$, then for all number fields K' containing K, the set $U(K')$ is finite.
2) For V a subvariety of an abelian variety, this broad conjecture is true and equivalent to the conjecture discussed in this paper. It is essentially the only case known. In particular, the conjecture is unknown even in the following "simple" example: let V be the quintic surface defined in $\mathbb{P}^3$ by the equation $X_0^5 + X_1^5 + X_2^5 + X_3^5 = 0$. Rational points are presumably not dense and concentrated on a finite number of curves like the lines $X_i + X_j = X_m + X_\ell = 0$. This is unknown, even in the function field case.

3 Diophantine equations over function fields, the "relative" case of Lang's conjecture

The main purpose of diophantine geometry is to describe sets $X(\mathbb{Q})$ of rational points of a variety. It is natural to look for generalizations, first to number fields then to fields finitely generated over $\mathbb{Q}$. The next step would be fields K finitely generated over $\mathbb{F}_p$. For example, some of the main results for X a curve are still true: if X is an elliptic curve, the group $X(K)$ is still finitely generated and if X has genus $g \geq 2$ then $X(K)$ is still finite.

Another motivation can be explained in a somewhat trivial case: if X is a curve defined over $\mathbb{Q}$, before trying to prove that $X(\mathbb{Q})$ is finite, one should perhaps check that $X(\mathbb{Q}(T)) \setminus X(\mathbb{Q})$ is finite. The deepest motivation however comes from the famous analogy between number fields and function fields studied by many mathematicians (let us quote arbitrarily Kronecker, Artin, Hasse, Weil, Néron, Grothendieck, Parshin, Arakelov,...). Let us give a formal definition:

Definition 3.1 *Let K_0 be an algebraically closed field, a function field K over K_0 (of transcendance degree 1) is the field of rational functions of a variety (of dimension 1) defined over K_0.*

This is actually the same as a finitely generated field over K_0 (of transcendance degree 1).
Now the natural question to ask is not "*is $X(K)$ finite?*" but "*is $X(K) \setminus X(K_0)$ finite?*". Let us give an example dear to number theorists.

Proposition 3.2 *Let X be the Fermat curve defined in the projective plane by $X^n + Y^n = Z^n$ where $n \geq 3$. Then $X(\mathbb{C}(T)) \setminus X(\mathbb{C})$ is empty.*

Proof: (see [Mas]) One actually proves a more general statement which easily implies the above. If A, B, C are non constant coprime polynomials with $A+B = C$ and if r, s, t are the number of distinct roots of A, B, C then

$$\max(deg(A), deg(B), deg(C)) \leq r + s + t - 1.$$

Indeed if X, Y, Z are coprime polynomials satisfying $X^n+Y^n = Z^n$, set $A = X^n$, $B = Y^n$ and $C = Z^n$. Applying the previous inequality and the observation that $deg(A) \geq nr$, $deg(B) \geq ns$ and $deg(C) \geq nt$, we obtain $n \max(r, s, t) \leq r + s + t - 1 \leq 3 \max(r, s, t) - 1$, hence $n \leq 2$.

In order to prove the inequality, one introduces the polynomial

$$\Delta := det \begin{pmatrix} A & A' \\ B & B' \end{pmatrix} = det \begin{pmatrix} A & A' \\ C & C' \end{pmatrix} = det \begin{pmatrix} C & C' \\ B & B' \end{pmatrix}.$$

This polynomial is non zero and has degree less than $deg(A) + deg(B) - 1$ (mutatis mutandis). If $A = a_0 \prod_{i=1}^{r} (T - a_i)^{\ell_i}$, $B = b_0 \prod_{i=1}^{s} (T - b_i)^{m_i}$ and $C = c_0 \prod_{i=1}^{t} (T - c_i)^{n_i}$ then $\prod_{i=1}^{r} (T - a_i)^{\ell_i - 1} \prod_{i=1}^{s} (T - b_i)^{m_i - 1} \prod_{i=1}^{t} (T - c_i)^{n_i - 1}$ divides Δ. Computing degrees gives the result. □

Since the proof uses derivations, it cannot be adapted to $\mathbb{Q}$!

We can look in this context for the analog of Mordell and Lang's conjecture. For example the following was shown in 1963 by Manin ([Man]):

Theorem 3.3 (Mordell's conjecture over function fields) *Let X be a curve of genus ≥ 2 defined over K a function field over K_0. Then $X(K)$ is finite unless X is isotrivial (which means that there is a curve X_0 defined over K_0 and isomorphic to X over some finite extension K' of K).*

A simpler but slightly incorrect statement is "$X(K)$ is finite unless X is defined over K_0". Manin proved this for $K_0 = \mathbb{C}$ (and hence for characteristic zero) and the corresponding characteristic p statement was proven later by Samuel.

The theorem proved by Hrushovski (Lang's conjecture for function fields) bears the same relation to the Lang conjecture as this last statement to the classical Mordell conjecture. Before stating it we review the analog of the Mordell-Weil theorem. We will need the notion of K/K_0-trace and K/K_0-image of an abelian variety defined over K (see Lang's book [Lan 59] on abelian varieties for details).

Proposition 3.4 1) *Let A be an abelian variety defined over K. There is an abelian variety A_0 defined over K_0 and a homomorphism (with finite kernel) $\tau : A_0 \to A$ such that for all B abelian variety defined over K_0 with a map $\tau' : B \to A$ we have a factorisation $\tau' = \tau \circ f$ via a morphism $f : B \to A_0$.*

Thus A_0 is the "biggest abelian subvariety of A defined over K_0", it is called the K/K_0-trace of A.
2) *Let A be an abelian variety defined over K. There is an abelian variety A_1 defined over K_0 and a surjective homomorphism $\pi : A \to A_1$ such that for all B abelian variety defined over K_0 with a map $\pi' : A \to B$ we have a factorisation $\pi' = f \circ \pi$ via a morphism $f : A_1 \to B$. Thus A_1 is the "biggest quotient of A defined over K_0", it is called the K/K_0-image of A.*

For our purposes it will clarify things to know that A_0 is zero if and only A_1 is zero (actually much more is true : A_0 and A_1 are "duals" hence have the same dimension). For a proof of the next theorem see [Lan 83].

Theorem 3.5 (Relative Mordell-Weil theorem, Lang-Néron) *Let A be an abelian variety defined over K and let $\tau : A_0 \to A$ be its K/K_0-trace, then the group $A(K)/\tau(A_0(K_0))$ is finitely generated.*

Remark: If the K/K_0-trace (or image) is zero then $A(K)$ is finitely generated. The translation of Lang's conjecture is easier to state when the stabilizer of the subvariety is finite (we saw that this is the crucial case). For a model theoretic translation see [Pi2].

Lang's conjecture over function fields *Let K be a function field over K_0 an algebraically closed field, let X be a subvariety of an abelian variety A both defined over K. Assume that* Stab_X *is finite. Let Γ be a subgroup of A of finite rank, defined over the algebraic closure of K, then either $X \cap \Gamma$ is not Zariski dense in X or there is a bijective morphism $X \to X_0$ onto a variety X_0 defined over K_0.*

Remark: If the K/K_0-trace of A is zero, then there can be no such X_0 (distinct from a point) because its Albanese variety would produce a non zero K/K_0-image for A. Hence in this case, the usual Lang conjecture is true.

4 Commented bibliography

Abelian and Jacobian varieties are treated in many books. For complex abelian varieties we especially quote as the easiest Swinnerton Dyer's book [Sw] and Rosen's paper in [CoSi], the book of Lange and Birkenhake [BiLa] being the most complete reference; it also includes Jacobians. The survey of Bost [Bos] contains also a lot of interesting material. The algebraic theory of abelian varieties is treated in Mumford's book [Mu] and Milne's first paper in [CoSi]. Milne's second paper in [CoSi] develops the algebraic theory of Jacobians; an exposition quite close to Weil's original treatment can be found in Serre's book [Se]. The books of Serge Lang provide also interesting different points of view [Lan 83], [Lan 72], [Lan 59].

One can find a quite thorough discussion of Lang's conjecture by Lang himself in [Lan 83], in his survey for the russian encyclopedia [Lan 91] as well as in his original papers [Lan 60], [Lan 65]. The Mordell conjecture was stated in Mordell's paper [Mo] in 1922 and first proven over function fields by Manin [Man] in 1963 and over number fields by Faltings [Fa 83] in 1983. The paper of Faltings [Fa 91], contains a proof that $X(k)$ is finite when k is a number field, X is a subvariety of an abelian variety and X does not contain a translate of a (non zero) abelian subvariety; his subsequent paper [Fa 94] deals with the case of a subvariety of an abelian variety. The methods rely heavily on ideas introduced by Vojta [Vo 91] who gave a completely new proof of Mordell's conjecture over number fields. This implies Lang's conjecture for finitely generated subgroups. The reduction of the general conjecture to this case had already been worked out in Hindry [Hi 88]. The Manin-Mumford conjecture was first proven by Raynaud [Ray1], [Ray2], [Ray3] and extended to general commutative algebraic groups by Hindry [Hi 88]. Extensions to semi-abelian varieties of Lang's conjecture have been worked out by Vojta [Vo 96] for a finitely generated subgroup, and by McQuillan [McQ] for the reduction from "finite rank" to "finitely generated" group. The case of subvarieties of a multiplicative group $\mathbb{G}_m^r$ was proven before in full generality by Laurent [Lau].

All these works deal with characteristic zero. As mentioned before, the statement over number fields implies the general statement in characteristic zero. Nevertheless specific proofs for function fields are of great interest. Buium (see [Bu 92], [Bu 93]) has given such a proof under the mild assumption that X is smooth. Abramovic and Voloch [AbVo] proved the characteristic p case under extra assumptions. It should therefore be noticed that the proof of the characteristic p case was unknown, in full generality, before Hrushovski [Hr 96]! Also note that Hrushovski in fact proves it for semi-abelian varieties. It is also possible to deduce the relative statement in characteristic zero from the corresponding statement in characteristic p (see [Hr] and the remarks in [Lan 91]).

References

[AbVo] D. Abramovic and F. Voloch, *Towards a proof of the Mordell-Lang conjecture in characteristic p*, International Math. Research Notices 2, (1992), 103-115.

[BiLa] C. Birkenhake and H. Lange, *Complex abelian varieties*, Springer, 1992.

[Bos] J.B. Bost, *Introduction to compact Riemann surfaces, jacobians and abelian varieties*, in Number Theory and Physics, Springer, 1993.

[Bous] E. Bouscaren, *Proof of the Mordell-Lang conjecture for function fields*, this volume.

[Bu 92] A. Buium, *Intersections in jet spaces and a conjecture of Serge Lang*, Annals of Math. 136 (1992), 583-593.

[Bu 93] A. Buium, *Effective bounds for geometric Lang conjecture*, Duke J. Math. 71 (1993), 475-499.

[CoSi] G. Cornell and J. Silverman (ed), *Arithmetic Geometry*, Springer, 1986.

[Fa 83] G. Faltings, *Endlichkeitssätze für abelsche Varietäten über Zahlkörpern*, Inv. Math. 73 (1983), 349-366.

[Fa 91] G. Faltings, *Diophantine approximation on abelian varieties*, Annals of Math. 133 (1991), 549-576.

[Fa 94] G. Faltings, *The general case of Lang's conjecture*, in Barsotti's Symposium in Algebraic geometry, Acad. Press, 1994, 175-182.

[Hi 88] M. Hindry, *Autour d'une conjecture de Serge Lang*, Inventiones Math. 94 (1988), 575-603.

[Hi 92] M. Hindry, *Sur les conjectures de Mordell et Lang (d'après Vojta, Faltings et Bombieri)*, Astérisque 209, 1992, 39-56.

[Hr 96] E. Hrushovski, *The Mordell-Lang conjecture for function fields*, J.AMS 9 (1996), 667-690.

[Hr] E. Hrushovski, *Proof of Manin's theorem by reduction to positive characteristic*, this volume.

[Lan 58] S. Lang, *Introduction to algebraic Geometry*, Interscience Tracts in pure and applied Mathematics, Interscience Publishers, New York, 1958.

[Lan 59] S. Lang, *Abelian varieties*, Interscience, New York, 1959.

[Lan 60] S. Lang, *Integral points on curves*, Publ. Math. IHES, 1960.

[Lan 65] S. Lang, *Division points on curves*, Ann. Mat. Pura Appl. LXX (1965), 229-234.

[Lan 72] S. Lang, *Algebraic functions and abelian functions*, Addison Wesley, 1972.

[Lan 83] S. Lang, *Fundamentals of Diophantine Geometry*, Springer, 1983.

[Lan 91] S. Lang, *Number Theory III: Diophantine Geometry*, Encycopedia of Mathematical Sciences, Springer, 1991.

[Lau] M. Laurent, *Equations diophantiennes exponentielles*, Inventiones Math. 78 (1984), 299-327.

[Man] Y. Manin, *Rational points of algebraic curves over function fields*, Isvetzia 27(1963), 1395-1440 (AMS Transl. Ser II 50 (1966), 189-234).

[Mas] R.C. Mason, *Diophantine equations over function fields*, London Math. Soc. L. N. 96, Cambridge U.P., 1984.

[McQ] M. McQuillan, *Division points on semi-abelian varieties*, Inventiones Math. 120 (1995), 143-159.

[Mo] L. Mordell, *On the rational solutions of the indeterminate equation of the 3rd and 4th degree*, Math. Proc. Cambridge Philos. Soc. 21 (1922), 179-192.

[Mu] D. Mumford, *Abelian varieties*, Oxford University Press, 1974.

[Pi1] A. Pillay, *Model theory of algebraically closed fields*, this volume.

[Pi2] A. Pillay, *The model-theoretic content of Lang's conjecture*, this volume.

[Ray1] M. Raynaud, *Courbes sur une variété abélienne et points de torsion*, Inventiones Math. 71 (1983), 207-233.

[Ray2] M. Raynaud, *Sous-variétés d'une variété abélienne et points de torsion*, in Arithmetic and Geometry (dedicated to Shafarevich), Birkhäuser, 1983.

[Ray3] M. Raynaud, *Around the Mordell conjecture for function fields and a conjecture of Serge Lang*, in Algebraic Geometry, MLN 1016, Springer, 1983, 1-19.

[Se] J.P. Serre, *Algebraic Groups and Class Fields*, Springer, 1988.

[Sh] I.R. Shafarevich, *Basic Algebraic Geometry*, Springer, 1977.

[Sw] H. Swinnerton-Dyer, *Introduction to the analytical theory of abelian varieties*, London M.S. LN 14, Cambridge University Press, 1974.

[Vo 91] P. Vojta, *Siegel's theorem in the compact case*, Annals of Math. 133 (1991), 509-548.

[Vo 96] P. Vojta, *Integral points on subvarieties of semi-abelian varieties*, Inventiones Math. 126 (1996), 133-181.

The model-theoretic content of Lang's conjecture

ANAND PILLAY

1 Introduction

The rest of this volume is dedicated to explaining Hrushovski's model-theoretic approach ([Hr 96]) to the geometric case of a conjecture of Lang. See [Hi] for a presentation of the conjecture and [Bous] for the proof. The purpose of this note is to point out that the use of model-theoretic and stability-theoretic methods should not be so surprising, as the *full* Lang conjecture itself is equivalent to a purely model-theoretic statement. The structure $(\mathbb{Q}, +, .)$ is wild (undecidable, definable sets have no "structure" etc.), as is the structure $(\mathbb{C}, +, .)$ with a predicate for the rationals. What comes out of the diophantine-type conjectures on the other hand is that *certain* enrichments of the structure $(\mathbb{C}, +, .)$ (more specifically expansions obtained by adding a predicate, not for $\mathbb{Q}$ itself, but rather for the $\mathbb{Q}$-points of certain algebraic groups) are *not* wild, in particular are stable.

It should be said that the results of the present paper are quite soft, and were already known to some other people (specifically Hrushovski). I was motivated to consider these issues by a question of Angus Macintyre regarding the stability of the field of complexes enriched by a predicate for the torsion points of some abelian variety.

2 Stability and structures of Lang-type

The following conjecture was formulated around the mid 1960's by S. Lang, as a general statement subsuming both the Mordell and Manin-Mumford conjectures.

Conjecture 2.1 [Lang] *Let A be a semi-abelian variety over $\mathbb{C}$, Γ a "finite rank" subgroup of A and X a subvariety of A. Then $X \cap \Gamma$ is a finite union of cosets (i.e. translates of subgroups).*

In 2.1, we identify A with the set of its $\mathbb{C}$-rational points. Similarly for X. It should first be mentioned that 2.1 has recently been proved: Faltings proved it for the case where A is an abelian variety, and Γ a finitely generated subgroup

Author partially supported by a grant from the NSF.

([Fa 94]). Vojta extended this to the case where A is a semi-abelian variety. Hindry showed how to replace "finitely generated" by "finite-rank" (see [Hi 88]). Before continuing we make some further remarks about 2.1 and its implications. By a semi-abelian variety we mean a connected algebraic group A such that some algebraic subgroup T of A is an algebraic torus (a product of a finite number of copies of the multiplicative group) and A/T is an abelian variety. Such a group is commutative and divisible. By a *finite rank* subgroup of A we mean a subgroup which is contained in the divisible hull of a finitely-generated subgroup. Suppose that A is an abelian variety defined over a number field F. The Mordell-Weil theorem asserts that $A(F)$ (the group of F-rational points of A) is finitely-generated. So the Lang conjecture implies that $X(F)$ is a finite union of cosets, whenever X is a subvariety of A, defined over F. This holds in particular when X is a smooth projective curve defined over F, the genus of X is at least 2, and A is the Jacobian variety of X. This forces $X(F)$ to be finite, yielding the Mordell conjecture. With the same notation, clearly $Tor(A)$ (the group of torsion points of A) is a finite rank subgroup of A, and thus the Lang conjecture implies that $X \cap Tor(A)$ is finite, yielding the Manin-Mumford conjecture.

The reader should look at Chapter I, section 6 of [Lan 91] for more information and background on the above matters.

Remark 2.2 *Suppose G is an algebraic group over an algebraically closed field K. Identify G with the group of its K-rational points. Let X be a subvariety of G and Γ be a subgroup of G. Then $X \cap \Gamma$ is a finite union of cosets if and only if there are connected algebraic subgroups $G_1, ..., G_n$ of G and translates C_i of the G_i such that $C_i \subseteq X$ for each i and $X \cap \Gamma \subseteq C_1 \cup ..... \cup C_n$*

Proof. Right implies left is trivial. The other direction follows by taking the Zariski closures of the relevant cosets. □

Definition 2.3 *Let K be an algebraically closed field, and A a commutative algebraic group over K (which we identify with its set of F-rational points). We say that the triple (K, A, Γ) is of* Lang-type *if for every $n < \omega$ and every subvariety X (over K) of A^n, $X \cap \Gamma^n$ is a finite union of cosets.*

(In the above A^n denotes the Cartesian product of A with itsef n times.)

Remark 2.4 *Thus Lang's conjecture says that if Γ is a finite rank subgroup of a semiabelian variety A over $\mathbb{C}$, then $(\mathbb{C}, A, \Gamma)$ is of Lang-type.*

Stability is an important property of a theory. We recall the "counting types" definition of stability: the complete theory T is stable if for any model M of T the set of complete types over M has cardinality at most $|M|^{|T|}$.(See [Zie] for an alternate definition.) The following is a convenient relativisation to a formula of the important notion "one-based".

Definition 2.5 *Let T be a complete stable theory and $\theta(x)$ a formula of the language of T. We say that $\theta(\bar{x})$ is* one-based *if whenever $\bar{a}$ is a tuple of realisations* *of θ in a model M of T, and A is a subset of M then $Cb(stp(\bar{a}/A)) \subseteq acl(\bar{a})$ (where acl is computed in M^{eq}).*

We will prove

Proposition 2.6 *Let K be an algebraically closed field, A a commutative algebraic group over K, and Γ a subgroup of A. Then (K, A, Γ) is of Lang-type if and only if $Th(K, +, ., \Gamma, a)_{a \in K}$ is stable, and the formula "$x \in \Gamma$" is one-based.*

A key result which immediately makes the connection between the one-based property and the Lang-type hypothesis is the following (a proof of which appears in [Las] in this volume):

Fact 2.7 ([HrPi]) *Let T be stable, M a big model of T and G an $\emptyset$-definable group in M. Then G is one-based in T if and only if every every definable (with parameters) subset of G^n is a finite Boolean combination of cosets (of definable subgroups of G^n).*

We now begin the proof of Proposition 2.6. The right to left direction is basically a direct consequence of 2.7. For suppose X to be a subvariety of A^n. $X \cap \Gamma^n$ is then a definable set in the structure $(K, +, ., \Gamma, a)_{a \in K}$. So by hypothesis and 2.7, $X \cap \Gamma^n$ is a finite Boolean combination of cosets. However it is an easy exercise to show that the Zariski closure (in a commutative algebraic group) of a finite Boolean combination of cosets is just a finite union of cosets of algebraic subgroups. So by the trivial direction of Remark 2.2, $X \cap \Gamma^n$ is a finite union of cosets.

We now prove the left to right direction making use of a couple of lemmas. Let us assume that the triple (K, A, Γ) is of Lang-type. Let M be the structure $(K, A, \Gamma, a)_{a \in K}$. Let $T = Th(M)$. It is convenient to let P denote the predicate picking out Γ in M. Note that our desired conclusion says something about *all* models of T, not just M.

Let us denote by $\mathbf{\Gamma}$ the structure $(\Gamma, X \cap \Gamma^n;\ X$ a subset of A^n definable with parameters in $(K, +, .))$.

Lemma 2.8 *$T_0 = Th(\mathbf{\Gamma})$ is one-based.*

Proof. Any subset X of A^n definable (with parameters) in $(K, +, .)$ is a finite Boolean combination of (irreducible) subvarieties X_i of A^n. Thus $X \cap \Gamma^n$ is a finite Boolean combination of the $X_i \cap \Gamma^n$. On the other hand the hypothesis that (K, A, Γ) is of Lang-type implies that for each i, $X_i \cap \Gamma^n$ is equal to $\cup_i (C_{ij} \cap \Gamma^n)$ where each C_{ij} is a translate of an algebraic subgroup A_{ij} of A^n. So clearly every set definable in the structure $\mathbf{\Gamma}$ is already definable with parameters in the reduct $\mathbf{\Gamma_0} = (\Gamma, +, B \cap \Gamma^n$: B an algebraic subgroup of $A)$. Now, essentially the proof of pp elimination in modules (suitably generalised as in [GuRe]) shows that every formula of $Th(\mathbf{\Gamma_0})$ is equivalent to a Boolean combination of positive primitive formulas. In particular, every definable set of n-tuples is a finite Boolean combination of cosets. Thus $Th(\mathbf{\Gamma_0})$ and so also $Th(\mathbf{\Gamma})$ is stable and one-based. □

We will prove that T is stable and the predicate P is one-based by the following argument. Let $\kappa > card(T)$ be chosen such that T_0 is κ-stable. Let N be a model of T which has cardinality λ and is λ-saturated for some $\lambda \geq \kappa$. N is an elementary extension of M of the form $(K_1, +, ., \Gamma_1, a)_{a \in K}$ where Γ_1 denotes the interpretation of P in N. Now Γ_1 can be equipped in a canonical manner with structure making it into an elementary extension $\mathbf{\Gamma_1}$ of the structure $\mathbf{\Gamma}$. (Namely we add predicates for $Y \cap \Gamma_1^n$ whenever Y is a subset of $(A(K_1))^n$ definable in the structure $(K_1, +, .)$ with parameters from K.) In the following $acl_f(\text{-})$ denotes algebraic closure in the field-theoretic sense.

Lemma 2.9 *Let B be a subset of N of cardinality at most κ. Then there is a subset C of Γ_1 of cardinality at most κ such that, whenever $\mathbf{a_1}$, $\mathbf{a_2}$ are tuples from Γ_1 with the same type over C in the structure $\mathbf{\Gamma_1}$, then there is a map $f : acl_f(B \cup K \cup \Gamma_1) \longrightarrow acl_f(B \cup K \cup \Gamma_1)$ such that*
(i) f fixes $B \cup K \cup C$ pointwise.
(ii) $f \mid \Gamma_1$ is an automorphism of the structure $\mathbf{\Gamma_1}$.
(iii) f is a partial elementary map in the sense of $(K_1, +, ., a)_{a \in K}$
(iv) $f(\mathbf{a_1}) = \mathbf{a_2}$
(v) for any $c, d \in N - acl_f(B \cup K \cup \Gamma_1)$ there is an automorphism g of N which extends f and takes c to d.

Proof of Lemma 2.9. We may assume that $K \subseteq B$. For each tuple b from B, let C_b be some finite subset of Γ_1 such that the type of b over Γ_1 in the structure $(K_1, +, ., a)_{a \in K}$ is definable over C_b. (This exists simply by ω-stability of $Th(K_1, +, .)$). Let $C = \bigcup\{C_b : b \text{ some tuple from } B\}$. So C has cardinality at most κ. Let $\mathbf{a_1}$, $\mathbf{a_2}$ be tuples from Γ_1 with the same type over C in the structure $\mathbf{\Gamma_1}$. Let f_1 be an automorphism of $\mathbf{\Gamma_1}$ which fixes C pointwise and takes $\mathbf{a_1}$ to $\mathbf{a_2}$. □

Claim For any tuple $\mathbf{b}$ from B, and any tuple $\mathbf{a}$ from $\mathbf{\Gamma_1}$, $(\mathbf{b}, \mathbf{a})$ and $(\mathbf{b}, \mathbf{f_1}(\mathbf{a}))$ have the same type in $(K_1, +, ., a)_{a \in K}$
Proof of claim. Let $\phi(\mathbf{x}, \mathbf{y})$ be a formula in the language of $(K_1, +, ., a)_{a \in K}$. By choice of C there is some tuple $\mathbf{c}$ from C and formula $\psi(\mathbf{y}, \mathbf{z})$ in the same language such that:
$K_1 \models \phi(\mathbf{b}, \mathbf{a})$ iff $K_1 \models \psi(\mathbf{a}, \mathbf{c})$.
However $\psi(\mathbf{y}, \mathbf{z}) \cap \mathbf{\Gamma_1}$ is a $\emptyset$-definable set in $\mathbf{\Gamma_1}$. So as f_1 is elementary from the point of view of $\mathbf{\Gamma_1}$, we see that
$K_1 \models \phi(\mathbf{b}, \mathbf{a})$ iff $K_1 \models \psi(f_1(\mathbf{a}), \mathbf{c})$. Thus $K_1 \models \phi(\mathbf{b}, \mathbf{a})$ iff $K_1 \models \phi(\mathbf{b}, \mathbf{f_1}(\mathbf{a}))$. This proves the claim. □

It follows from the claim that if we let f_2 be the map from $B \cup \Gamma_1$ to itself which extends f_1 and is the identity on B, then f_2 is elementary in the sense of $(K_1, +, ., a)_{a \in K}$. Let f be some extension of f_2 to an elementary (in the sense of $(K_1, +, ., a)_{a \in K}$) permutation of $acl_f(B \cup \Gamma_1)$. f satisfies (i), (ii), (iii) and (iv) of the lemma. For any $c, d \in N - acl_f(B \cup \Gamma_1)$, f clearly extends to an automorphism g of the structure $(K_1, +, ., a)_{a \in K}$, which takes c to d. As g fixes Γ_1 setwise, g is also an automorphism of the structure N. The lemma is proved. From the above lemma we conclude two things. First, as T_0 is κ-stable, so is

T. Thus T is stable. Secondly, if $N = (K_1, +, ., \Gamma_1, a)_{a\in K}$ is a model of T, then any subset of Γ_1^n definable in N is already definable in $\mathbf{\Gamma_1}$. Thus from Fact 2.7 and Lemma 2.8, the predicate P is one-based for T.
We have completed the proof of the left to right direction of 2.6. □

3 Concluding remarks.

The truth of Lang's conjecture together with Proposition 2.6 gives us a number of "new" stable theories of "arithmetic-type", namely $Th(\mathbb{C}, +, ., \Gamma)$ whenever Γ is a finite-rank subgroup of a semi-abelian variety over $\mathbb{C}$. Some special cases of the latter (for example where Γ is the group of complex roots of unity) were already observed by Zilber, using a number-theoretic result due to Mann, together with arguments like those in the proof of 2.6. Actually the case in which Γ is a finitely generated subgroup of $(acl(\mathbb{Q}))^*)^n$ follows from Schmidt's Subspace Theorem (a result in diophantine approximation), as pointed out in [Ev].

Bruno Poizat pointed out to me that the proof of Proposition 2.6 really yields information about arbitrary strongly minimal structures.

Proposition 3.1 *Suppose D is a strongly minimal set, and Γ is an arbitrary subset of D^n. Let $\mathbf{\Gamma}$ be the structure whose universe is Γ and whose relations are the sets $X \cap \Gamma^m$ for X a definable (with parameters) subset of D^{nm} (with m varying). Let $\mathbf{\Gamma}'$ be the structure whose universe is Γ and whose relations are those subsets of Γ^m (m varying) definable (with parameters) in the structure (D, Γ). Then $\mathbf{\Gamma}$ and $\mathbf{\Gamma}'$ have the same $\emptyset$-definable relations.*

References

[Bous] E. Bouscaren, *Proof of the Mordell-Lang conjecture for function fields*, this volume.

[Ev] J.H. Evertse, *The Subspace Theorem of W.M.Schmidt*, in Diophantine Approximation and Abelian Varieties, Springer Lecture Notes 1566, 1993, 31-50.

[Fa 94] G. Faltings, *The general case of Lang's conjecture*, in Barsotti's Symposium in Algebraic geometry, Acad. Press, 1994, 175-182.

[GuRe] H.B. Gute and K.K. Reiter, *The last word on quantifier elimination in modules*, Journal of Symbolic Logic 55 (1990), 670-673.

[Hi 88] M. Hindry, *Autour d'une conjecture de Serge Lang*, Inventiones Math. 94 (1988), 575-603.

[Hi] M. Hindry, *Introduction to abelian varieties and the Lang Conjecture*, this volume.

[Hr 96] E. Hrushovski, *The Mordell-Lang conjecture for function fields*, J.AMS 9 (1996), 667-690.

[HrPi] E. Hrushovski and A. Pillay, ***Weakly normal groups***, Logic Colloquium '85, North-Holland, 1987, 233-244.

[Lan 91] S. Lang, *Number Theory III: Diophantine Geometry*, Encycopedia of Mathematical Sciences, Springer, 1991.

[Las] D. Lascar, ***ω-stable groups***, this volume.

[Zie] M. Ziegler, ***Introduction to stability theory and Morley rank***, this volume.

Zariski geometries

DAVID MARKER

Zariski geometries were introduced by Hrushovski and Zilber in [HrZi 96], [HrZi 93] and [Zil]. From a technical point of view this work provides a class of strongly minimal sets where Zilber's conjecture holds (see [Zie, end of section 5]. It also provides the answer to two metamathematical questions. ***How do you characterize the topological spaces that arise from the Zariski topology of an algebraic curve? Can you recover the field from the topological spaces?*** The answer to these questions is provided by Theorem 3.3 below. This result plays a key role in Hrushovski's proof of the Mordell-Lang conjecture for function fields.

In these notes I will try to give careful statements of the main definitions and theorems about Zariski geometries and discuss the proof from [HrSo] that strongly minimal sets in differentially closed fields of characteristic zero are (after perhaps discarding finitely many points) Zariski geometries. These results will play a crucial role in [Bous] in the proof of Hrushovski's theorem. I will also try to give some vague ideas about the methods used in the proof and give geometric examples which illustrate the main technical ideas. Finally in §5 I will present briefly a generalization of the notion of Zariski geometry which is needed in Hrushovski's proof in positive characteristic (see [De]).

I am grateful to Anand Pillay for many discussions on these topics and Elisabeth Bouscaren for many comments on a preliminary version of these notes.

1 Definitions

Let X be a topological space. We say that the topology on X is *Noetherian* if there is no infinite descending sequence of closed sets. For example:

i) Let K be a field and consider the Zariski topology on K^n. A descending sequence of closed sets corresponds to an ascending sequence of ideals in $K[X_1, \ldots, X_n]$. By Hilbert's Basis Theorem any such sequence is finite.

ii) Let K be a differentially closed field (see [Wo] for relevant definitions). We say that $C \subseteq K^n$ is δ-closed if C is a finite union of sets of the form

Author partially supported by NSF grants DMS-9306159, DMS-9626856 and INT-9224546, and an AMS Centennial Fellowship.

$\{\overline{x} \in K^n : \bigwedge p_i(\overline{x}) = 0\}$ where $p_1, \ldots, p_n$ are differential polynomials. If $C \subset K^n$ is δ-closed then

$$I^\delta(C) = \{p \in K\{X_1, \ldots, X_n\} : \forall a \in C \; p(a) = 0\}$$

is a radical differential ideal. A descending sequence of closed sets corresponds to an ascending sequence of radical differential ideals and Ritt's basis theorem ([Mar 96] 1.16) insures that such a sequence is finite.

iii) Let α be an ordinal and let $X = \alpha$. Topologize X such that the closed sets are those of the form $F_\beta = \{\gamma : \gamma < \beta\}$ for $\beta \leq \alpha$. This is a Noetherian topology.

If $C \subseteq X$ is closed, we say that C is *irreducible* if whenever F_0 and F_1 are closed and $C = F_0 \cup F_1$, then $C = F_0$ or $C = F_1$.

Lemma 1.1 *Let X be a Noetherian space and let $C \subseteq X$ be closed. There are irreducible closed sets $C_1, \ldots, C_n$ such that $C = C_1 \cup \ldots \cup C_n$. Moreover, if we choose $C_1, \ldots, C_n$ such that $C_i \not\subseteq C_j$ for $i \neq j$, then $C_1, \ldots, C_n$ are unique up to permutation.*

We call $C_1, \ldots, C_n$ the *irreducible components* of C.

The proof is left as an exercise. [Hint: To show that there is such a decomposition, suppose not and use the reducibility of C to build an infinite finite branching tree of closed subsets of C. For uniqueness, first note that if $B \subset C$ is irreducible, then $B \subseteq C_i$ for some i.]

If X is a Noetherian space and $C \subseteq X$ is irreducible, closed, and nonempty, then we inductively define the *dimension* of C as follows:

$dim\ C \geq 0$;

$dim\ C = \sup \{dim\ F + 1 : F \subset C,\ F$ closed, irreducible, and nonempty$\}$.

If X is an arbitrary closed set the dimension of X is the maximum dimension of its irreducible components.

If K is an algebraically closed field, and $C \subseteq K^n$ is Zariski closed and irreducible, then $dim\ C$ is the transcendence degree of the coordinate ring $K[X_1, \ldots, X_n]/I(C)$ over K. Thus $dim\ C \leq n$.

In general, the dimension of a Noetherian space is an ordinal. It is easy to see that in example iii) $dim\ F_{n+1} = n$ and $dim\ F_\beta = \beta$ for $\beta \geq \omega$.

If K is a differentially closed field and $C \subset K$ is irreducible and δ-closed, then $dim\ C$ is less than or equal to the transcendence degree over K of the differential coordinate ring $K[X, X', X'', \ldots]/I^\delta(C)$ (see [Mar 96] 5.4). Thus $dim\ C \leq \omega$. On the other hand if $C^n = \{x \in K : x^{(n)} = 0\}$, then $C_1 \subset C_2 \subset \ldots$. Thus $dim\ K = \omega$.

A *Zariski geometry* is an infinite set D together with a family of Noetherian topologies on $D, D^2, D^3, \ldots$ such that the following axioms hold:

(Z0) [Coherence and Separation]: i) If $f : D^n \to D^m$ is defined by $f(x) = (f_1(x), \ldots, f_n(x))$ where each $f_i : D^n \to D$ is either constant or a coordinate projection, then f is continuous.

ii) Each diagonal $\Delta^n_{i,j} = \{x \in D^n : x_i = x_j\}$ is closed.

(Z1) [Weak QE]: If $C \subseteq D^n$ is closed and irreducible, and $\pi : D^n \to D^m$ is a projection, then there is a closed $F \subset \overline{\pi(C)}$ such that $\pi(C) \supseteq \overline{\pi(C)} \setminus F$.

(Z2) [Uniform one-dimensionality]: i) D is irreducible.

ii) Let $C \subset D^n \times D$ be closed and irreducible. For $\overline{a} \in D^n$, let $C(\overline{a}) = \{x \in D : (\overline{a}, x) \in C\}$. There is a number N such that for all $\overline{a} \in D^n$, either $|C(\overline{a})| \leq N$ or $C(\overline{a}) = D$. In particular any proper closed subset of D is finite.

(Z3) [Dimension theorem]: Let $C \subseteq D^n$ be closed and irreducible. Let W be a non-empty irreducible component of $C \cap \Delta^n_{i,j}$. Then $dim\ C \leq dim\ W + 1$.

Remarks:

1) (Z0) insures that the topology on D^n refines the product topology. If $C \subset D^n \times D^m$ is closed, and $\overline{a} \in D^n$, then $C(\overline{a}) = f^{-1}C$, where $f(x) = (\overline{a}, \overline{x})$, for $\overline{x} \in D^m$. Thus $C(\overline{a})$ is closed. Also, if $a \in D$, then $f(x) = (a, x)$ is a continuous function from D to D^2. Since the diagonal of D^2 is closed, $f^{-1}(\Delta^2_{1,2}) = \{a\}$ is closed. Thus singletons are closed.

2) We say that a Zariski geometry is *complete* if whenever $C \subseteq D^n$ is closed and $\pi : D^n \to D^m$ is a projection, then $\pi(C)$ is closed. Completeness is a strong form of (Z1).

3) It can be shown that

$$dim\ C_0 \times C_1 = dim\ C_0 + dim\ C_1.$$

In particular, $dim\ D^n = n$. Once we know that every set has finite dimension, (Z3) has this more natural statement: if $C \subset D^n$ is closed and irreducible and W is a non-empty irreducible component of $C \cap \Delta^n_{i,j}$, then $dim\ W \geq dim\ C - 1$.

4) Suppose C_0, C_1 are closed irreducible subsets of D^n. Consider $I = \{(x, y) \in D^{2n} : x \in C_0, y \in C_1, x = y\}$. An easy induction applying (Z3) to $C_0 \times C_1$ shows that if W is an irreducible component of I, then $dim\ W \geq dim\ C_0 + dim\ C_1 - n$.

Suppose D is an irreducible (quasi-projective) algebraic curve over an algebraically closed field (we refer to [Pi1] for relevant definitions). Topologize D^n by taking the restriction of the Zariski topology.

(Z0) holds trivially.

(Z1) Suppose $C \subset D^n$ is closed and $\pi : D^n \to D^m$ is a projection. By quantifier elimination for algebraically closed fields, $\pi(C) = \bigcup F_i \setminus E_i$, where $F_i \subseteq D^m$ is closed and irreducible and $E_i \subset F_i$ is closed. Then $\overline{\pi(C)} = \bigcup F_i$ and if $E = \bigcup E_i$, then $\pi(C) \supseteq \bigcup F_i \setminus E$.

For projective curves, the completeness of projective varieties (see [Sha] I.5.2) implies that D is complete in the sense defined above.

(Z2) Since D is irreducible, D is strongly minimal. Thus for any definable $C \subseteq D^n \times D$, there is an N such that $|C(\overline{a})| \leq N$ or $|D \setminus C(\overline{a})| \leq N$ for all $\overline{a} \in D^n$. If C is closed, then $C(\overline{a})$ is closed; thus if $C(\overline{a})$ is infinite, then $C(\overline{a}) = D$.

(Z3) is more problematic. If D is a smooth curve, then D^n is a smooth variety and (Z3) follows from the next result (see [Da] 6.5 or the appendix).

Lemma 1.2 *Suppose V is a smooth variety over an algebraically closed field, and both C_0 and C_1 are irreducible subvarieties. If W is a non-empty irreducible component of $C_0 \cap C_1$, then* dim $W \geq$ dim C_0 + dim C_1 − dim V.

Corollary 1.3 *Let D be a smooth quasi-projective algebraic curve, then D, equipped with the Zariski topologies on D^n, is a Zariski geometry.*

We next give an example showing that (Z3) does not hold for arbitrary algebraic curves. Let D be the projective curve with homogeneous equation

$$Y^2Z = X^3 + X^2Z.$$

Making the usual identification of the affine plane $\mathbb{A}^2$ in $\mathbb{P}^2$, D is the closure of C, where C is the affine singular cubic $y^2 = x^3 + x^2$. Define $f : \mathbb{P}^1 \to D$, the desingularization of D, by

$$(U,V) \mapsto (U^2V - V^3, U^3 - UV^2, V^3).$$

As usual we have the affine line $\mathbb{A}^1$ embedded into $\mathbb{P}^1$ by $t \mapsto (t,1)$. Then $f|\mathbb{A}^1$ maps $\mathbb{A}^1$ onto C by $t \mapsto (t^2-1, t(t^2-1))$ and f maps the point at infinity of $\mathbb{P}^1$ to the point at infinity of D. The morphism f maps $\mathbb{P}^1$ onto D, and is almost one-to-one except that $f(1,1) = f(-1,1) = (0,0,1)$.

Define $g : \mathbb{P}^1 \times \mathbb{P}^1 \times \mathbb{P}^1 \to D^3$ by $g(x,y,z) = (f(x), f(y), f(z))$. We view $\mathbb{A}^3$ as a dense subset of $\mathbb{P}^1 \times \mathbb{P}^1 \times \mathbb{P}^1$ and note that $\mathbb{A}^3 = g^{-1}(C^3)$. Let $F_0, F_1 \subset \mathbb{P}^1 \times \mathbb{P}^1 \times \mathbb{P}^1$ be the closures of

$$x + y + z = 3$$

and

$$x + y + z = -3,$$

respectively. Let E_i be the image of F_i under g. By the completeness of projective varieties, E_i is closed, and it is clearly irreducible of dimension 2. All of the irreducible components of $F_0 \cap F_1$ are contained in $\mathbb{P}^1 \times \mathbb{P}^1 \times \mathbb{P}^1 \setminus \mathbb{A}^3$, while $E_0 \cap E_1 \cap C^3 = \{(0,0,0)\}$. Thus

$$E_0 \cap E_1 \subseteq D^3 \setminus C^3 \cup \{(0,0,0)\}.$$

Since $D^3 \setminus C^3 = \{(a,b,c) \in D^3 : a = \infty \vee b = \infty \vee c = \infty\}$, where ∞ is the unique point at infinity of D, $D^3 \setminus C^3$ is closed. Thus $\{(0,0,0)\}$ is an irreducible component of $E_0 \cap E_1$ and (Z3) fails for D.

On the other hand, nonsingularity is not necessary for (Z3). Let D be the singular projective curve with homogeneous equation $Y^2Z = X^3$. The desingularization $(U,V) \mapsto (U^2V, U^3, V^3)$ is one-to -one. It follows that the dimension theorem holds for D.

The trichotomy of strongly minimal sets introduced here in [Zie] arises in the context of Zariski geometries. Algebraically closed fields with the Zariski topology are non-locally modular Zariski geometries. It is easy to build trivial and non-trivial locally modular examples. If D is an infinite set and the closed subsets of D^n are positive Boolean combinations of formulas $x_i = x_j$ and $x_i = a$ for $a \in D$, then D is a trivial Zariski geometry. If D is a one-dimensional vector space over a division ring and the closed sets of D^n are finite unions of cosets of subspaces, then D is a locally modular Zariski geometry. In §3 we will examine non-locally modular Zariski geometries.

2 Preliminary concerns

Let D be a Zariski geometry. Let $\mathcal{L}_D$ be the language where we have an n-ary predicate for each closed subset of D^n. Let T_D be the $\mathcal{L}_D$-theory of D. Properties (Z1) and (Z2) have natural model theoretic analogs. We say that $X \subseteq D^n$ is *constructible* if it is a finite Boolean combination of closed sets. The reader should consult [Zie] for unexplained model theoretic ideas. Note that we use the notation MR for Morley rank.

Theorem 2.1 [HrZi 96] *i) T_D admits elimination of quantifiers. In other words, the projection of a constructible set is constructible.*

ii) T_D is ω-stable of finite Morley rank and if C is constructible, then MR $C = \dim \overline{C}$. *In particular D is strongly minimal.*

Note that we have added a predicate symbol for each element of D. Thus $D = \mathrm{dcl}(\emptyset)$.

Remarkably, if M is an elementary extension of D, then M itself can be viewed as a Zariski geometry. Our basic closed sets are those X for which there is a closed $C \subseteq D^m \times D^n$ and $\overline{a} \in M^m$ such that $X = C(\overline{a})$, i.e.

$$X = \{\overline{b} \in M^n : M \models C(\overline{a}, \overline{b})\}.$$

Theorem 2.2 [HrZi 96] *The induced topology on M^n is Noetherian and, with respect to these topologies, M is a Zariski geometry.*

As we will be primarily interested in questions about interpretability, we can replace D by a saturated elementary extension. Thus without loss of generality we may assume that there is a Zariski geometry D_0 such that D is a saturated elementary extension of D_0 in the language $\mathcal{L}_{D_0}$ and the geometry on D is the natural extension of D_0. **Henceforth we will assume that D is saturated in this sense.**

If $A \subset D$, we say that X is A-closed if $X = C(\overline{a})$ for some C in $\mathcal{L}_{D_0}$ and $\overline{a} \in A$. If $C \subseteq D^n$ is irreducible and A-closed, we say that $\overline{a} \in C$ is *A-generic*, if $\overline{a} \notin F$ for any A-closed $F \subset C$. We will often be cavalier and say "generic" when we mean A-generic for suitable A. If C is irreducible and $A \subseteq C$ is constructible, then we say A is generic iff it contains a generic of C iff $dim\ \overline{C \setminus A} < dim\ C$.

If $a \in D$, we define the ***locus*** of a over A to be the smallest A-closed set containing a.

Theorem 2.1 and the rank inequalities have the following useful geometric consequence.

Corollary 2.3 (Generic fibers lemma) *Let* $C \subseteq D^{n+m}$ *and let* $\pi : D^n \times D^m \to D^m$ *be a projection. If* $\overline{a} \in \overline{\pi C}$ *is generic, then*

$$dim\ C = dim\ \overline{\pi C} + dim\ \pi^{-1}(\overline{a}).$$

In general, Zariski geometries will not have elimination of imaginaries (see [Pil] or [Zie]). Let K be algebraically closed and let C be a smooth projective curve of genus at least one. We claim that C, viewed as a Zariski geometry, does not eliminate imaginaries. There is a non-trivial projection π from C onto $\mathbb{P}^1$ which induces a definable equivalence relation on C with quotient $\mathbb{P}^1$; namely $x\ E\ y$ if and only if $\pi(x) = \pi(y)$. If we could eliminate imaginaries in C, then there would be a rational map from a Zariski open subset of $\mathbb{P}^1$ in C^n. Taking a suitable projection from C^n onto C, we would get a rational map from an open subset of $\mathbb{P}^1$ onto C. This extends to a morphism from $\mathbb{P}^1$ onto C (see [Ha] pg.43). But this is impossible since C has nonzero genus (this follows, for example, from Hurwitz theorem, see [Ha] pg.299). A more extreme example is given in §6.

Fortunately we need only to consider simple imaginary elements.

We say that S is a ***special sort*** if there is a natural number n and H a subgroup of Sym(n) such that $S = D^n/H$, where we view Sym(n) as acting on D^n by permuting coordinates. We use the following lemma of Lascar and Pillay (for proof see [Pil] 1.6).

Lemma 2.4 *If* D *is a strongly minimal set and the algebraic closure of* $\emptyset$ *is infinite, then every element of* D^{eq} *is interdefinable with an element of a special sort.*

In Zariski geometries, $acl(\emptyset)$ is the prime model of T_D; thus we can eliminate imaginaries down to special sorts. If $S = D^n/H$, then there is a natural finite-to-one projection of D^n onto S. We topologize S such that $X \subseteq S$ is closed if and only if its inverse image is closed. We next isolate a collection of subsets of special sorts where there is a good intersection theory.

Let $C \subseteq D^n$ be an irreducible closed set and $p \in C$. Let $\Delta_C = \{(x, y) \in C \times C : x = y\}$. We say that p is a ***regular*** point of C if there is a closed irreducible $G \subseteq D^n \times D^n$ such that

i) $codim\ G = dim\ C$, and

ii) Δ_C is the unique component of $G \cap C \times C$ passing through (p,p).

If $C \subset D^n$ is a reducible closed set and $p \in C$, then p is a regular point of C if and only if it is contained in a unique irreducible component of C and it is a regular point of that component.

If C is a closed subset of a special sort $S = D^n/H$, then $p \in C$ is *regular* if and only if for any $q \in \pi^{-1}(p)$, q is contained in a unique irreducible component of $\pi^{-1}(C)$ and q is a regular point of this component.

We say that a subset U of a special sort is regular if U is an open subset of $\overline{U}$ and every point of U is regular in $\overline{U}$. Any special sort is regular.

Lemma 2.5 [HrZi 96] *i) If C is a closed subset of a special sort, then the set of regular points of C is a dense open subset.*

ii) If $X_1, \ldots, X_n$ are regular subsets of special sorts, and C_1, C_2 are closed subsets of $\prod X_i$, and W is a non-empty irreducible component of $C_1 \cap C_2$, then $\dim W \geq \dim C_1 + \dim C_2 - \sum \dim X_i$.

While the definition is non-intuitive, we will show that if V is an algebraic variety then every non-singular point is regular in the sense defined above.

To illustrate the idea we first sketch the proof for plane curves. The general case is similar but the notation is messy.

Let C be the curve $F(X,Y) = 0$ and suppose $p = (x_0, y_0)$ and $\frac{\partial F}{\partial Y}(x_0, y_0) \neq 0$. Let $G = \{(X,Y,Z,W) : X = Z\}$. It is easy to see that $G \cap (C \times C)$ has dimension 1, and thus Δ_C is an irreducible component. We claim that Δ_C is the unique component containing (p,p). It suffices to show that (p,p) is a non-singular point of $G \cap (C \times C)$, as any point on two components is singular. The (possibly reducible) algebraic set $G \cap (C \times C)$ is given by the equations:

$$F(X,Y) = 0$$

$$F(Z,W) = 0$$

and

$$X - Z = 0.$$

The Jacobian matrix at (p,p) is

$$J = \begin{pmatrix} \frac{\partial F}{\partial X}(x_0,y_0) & \frac{\partial F}{\partial Y}(x_0,y_0) & 0 & 0 \\ 0 & 0 & \frac{\partial F}{\partial X}(x_0,y_0) & \frac{\partial F}{\partial Y}(x_0,y_0) \\ 1 & 0 & -1 & 0 \end{pmatrix}.$$

Since $\frac{\partial F}{\partial Y}(x_0, y_0) \neq 0$, one sees that the rows are linearly independent. Thus the tangent space to $G \cap (C \times C)$ at (p,p) has dimension 1 as desired.

Let $V \subseteq K^n$ have dimension m. Suppose V is defined by the equations $F_1(\overline{X}) = \ldots = F_l(\overline{X}) = 0$. If $p = \overline{x}$ is a smooth point of V, then the matrix

$$J = \left(\frac{\partial F_i}{\partial X_j}(\overline{x})\right)$$

has rank $n-m$.

Renumbering equations and variables if necessary, assume that the minor

$$\left(\frac{\partial F_i}{\partial X_j}(\overline{x})\right) \quad i=1,\ldots,n-m,\ j=m+1,\ldots,n$$

is a nonsingular matrix.

Let $G=\{(\overline{x},\overline{y})\in K^{2n} : x_i=y_i \text{ for } i=1,\ldots,m\}$. As above, Δ_V is an irreducible component of $G\cap(V\times V)$. We argue that (p,p) is a nonsingular point of $G\cap(V\times V)$.

We must consider the $(2l+m)\times 2n$ matrix J where:

For $i\le l$, the ith row of J is

$$\left(\frac{\partial F_i}{\partial X_1}(\overline{x})\ldots\frac{\partial F_i}{\partial X_n}(\overline{x})\ 0\ldots 0\right)$$

and the $l+i^{th}$ row is

$$\left(0\ldots 0\ \frac{\partial F_i}{\partial X_1}(\overline{x})\ldots\frac{\partial F_i}{\partial X_n}(\overline{x})\right).$$

For $i\le m$ the $(2l+i)^{th}$ row has 1 in the ith column and -1 in the $n+i^{th}$ column. It is easy to see that J has rank $2(n-m)+m$. Thus, the tangent space at (p,p) has dimension $2n-(2n-m)=m$, as desired.

Finally, we define manifolds which play the role of connected abstract varieties.

Let D be a Zariski geometry. A *manifold* is a set X and a finite number of injective maps $g_i : U_i\to X$, where:

i) $\bigcup g_i(U_i)=X$;

ii) U_i is an irreducible regular subset of a special sort;

iii) for each i,j, $\{(x,y)\in U_i\times U_j : g_i(x)=g_j(y)\}$ is a closed irreducible subset of $U_i\times U_j$, projecting onto a nonempty open set $V_{i,j}\subseteq U_i$.

If M and N are manifolds, then $f : M\to N$ is a *morphism* if its graph is a closed irreducible subset of $M\times N$.

Using Lemma 2.5 it is easy to see that the dimension theorem holds for manifolds. The following theorem is proven by following Hrushovski's proof of that a group definable in an algebraically closed field is an algebraic group (see [Pil] 4.12).

Theorem 2.6 ([HrZi 96]) *If D is a Zariski geometry and G is an interpretable group then G can be definably topologized so that G is a manifold and multiplication is morphism. Similarly, if K is an interpretable field, then K can be given a manifold structure such that addition and multiplication are morphisms.*

3 Ample geometries

Let D be a saturated Zariski geometry.

A *plane curve* is a one-dimensional irreducible closed subset of D^2. A *family of plane curves* is given by E, an irreducible closed subset of a manifold, and $C \subset E \times D^2$, such that for generic $e \in E$, $C(e)$ is a plane curve.

We say that a family of plane curves is *ample* if there is a plane curve $C(e)$ with $p, q \in C(e)$ whenever p and q are independent generic points of D^2. An ample family is called *very ample* if for p, q (not necessarily independent) generic points in D^2 there is a curve $C(e)$ with $p \in C(e)$ and $q \notin C(e)$. In this case, we say that the family *separates points.*

We say that a Zariski geometry D is (very) ample if there is a (very) ample family of curves on D.

Example: Let D be an algebraic (quasi-projective) curve. Let $\phi : D^2 \to \mathbb{P}^n$ be an embedding. Assume that n is minimal. Let $X = \phi(D^2)$. Then X is a two dimensional variety. Let $Gr = \{h \subset \mathbb{P}^n : h \text{ is a hyperplane}\}$. Then Gr is an n-dimensional projective variety. Let $C = \{(h, x) : h \in Gr \text{ and } x \in \phi^{-1}(h)\}$.

Since n is minimal, for any $h \in Gr$, $h \cap X$ has dimension at most one and if h is generic, then $h \cap X$ is irreducible and one dimensional ([Ha] pg 179). Since ϕ is one-to-one, $C(h)$ is irreducible and one dimensional for generic h. For $x \in D^2$ it is easy to find a curve $C(h)$ with $x \in C(h)$. Moreover, for any $x, y \in D^2$ there is an h such that $x \in C(h)$ and $y \notin C(h)$.

Since Gr can be given a manifold structure, the Zariski geometry of any smooth curve is very ample. The main objective of [HrZi 96] is to see to what extent the converse of this is true.

We first show that Zariski geometries are non-locally modular if and only if they are ample. We recall the characterization of non-local modularity given in [Zie] 5.12.

Lemma 3.1 *A strongly minimal set is locally modular if and only if every family of plane curves is one dimensional.*

Suppose $C \subset E \times D^2$ is an ample Zariski geometry and $p, q \in D^2$ are independent generic points. Then there is $e \in E$ such that $p, q \in C(e)$. Since $C(e)$ has rank one, MR $(p, q/e) \leq 2$. Since

$$2 = \text{MR } (p, q) \leq \text{MR } (p, q, e) = \text{MR } (p, q/e) + \text{MR } (e),$$

MR $(e) \geq 2$. Thus D is non-locally modular.

Conversely, suppose D is a non-locally modular Zariski geometry. Let X be a strongly minimal set in D^2 with canonical base c such that MR $(c) \geq 2$ (i.e. MR $(tp(c/\emptyset) \geq 2)$). Without loss of generality assume that X is closed and irreducible. Let (a_0, b_0) be a generic point of X Since we can eliminate imaginaries using special sorts, there is an $e_0 \in D^n$ and a subgroup H of $\text{Sym}(n)$

such that $c = e_0/H$. Let C be the locus of (e_0, a_0, b_0) over $\emptyset$ and let E be the locus of e_0 over the empty set.

claim 1 If $e \in E$ is generic, then $C(e)$ is irreducible and one-dimensional.

This follows since e and e_0 realize the same type over the empty set.

claim 2 Let $\pi_2 : D^n \times D^2 \to D^2$ be the projection onto the last two coordinates. Then $\pi_2 C$ is dense in D^2.

It is easy to see that $\overline{\pi_2 C}$ is irreducible and $X \subseteq \pi_2(C)$. If $dim\ \pi_2 C < 2$, then $\pi_2 C = X$ and X is $\emptyset$-definable, a contradiction.

claim 3 $dim\ C \geq 3$, thus if $(a, b) \in D^2$ is generic and $E_{a,b} = \{e \in E : (e, a, b) \in C\}$, then MR $E_{a,b} \geq 1$. Also there is $e \in E_{a,b}$ such that $C(e)$ is irreducible and one-dimensional.

Let $\pi_1 : D^n \times D^2 \to D^n$ be the projection onto the first n-coordinates. By the generic fibers lemma, $dim\ C = dim\ E + dim\ C(e)$, where $e \in E$ is generic. Since $dim\ E \geq 2$ and $dim\ C(e) = 1$ for e generic, $dim\ C \geq 3$. Again by the generic fibers lemma, for $(a, b) \in D^2$ generic,

$$\text{MR } E_{a,b} = dim\ C - dim\ D^2 \geq 1.$$

Since $C(e_0)$ is irreducible and one-dimensional and (a_0, b_0) is a generic point of D^2, there must be $e \in E_{a,b}$ such that $C(e)$ is irreducible and one-dimensional (indeed e is a generic point of E over $\emptyset$).

claim 4 Let $I = \{(e, a, b, c, d) : e \in E, \text{ and } (a, b), (c, d) \in C(e)\}$. Then MR $I =$ MR $E + 2$. The projection of I onto D^4 has dimension 4.

Let e, a, b be a generic of C. Let $(c, d) \in C(e)$ be generic over e, a, b. Let W be the irreducible component of $E_{a,b}$ containing e. Then $dim\ W \geq 1$. Suppose $e_1 \in W$ and MR $(e_1/a, b, c, d) = dim\ W$. If $(c, d) \in C(e_1)$, then, by the genericity of e_1, for all $e_2 \in W$, $(c, d) \in C(e_2)$. By the genericity of (c, d),

$$X = \bigcup_{e' \in W} C(e')$$

a contradiction since for independent generic $e_1, e_2 \in W$, $C(e_1) \cap C(e_2)$ is finite. Thus MR $(e/a, b, c, d) <$ MR $(e/a, b)$. By symmetry of forking, MR $(c, d/a, b, e) <$ MR $(c, d/a, b)$. In other words,

$$\text{MR } (a, b, c, d, e) - \text{MR } (a, b, c, d) < \text{MR } (e, a, b) - \text{MR } (a, b).$$

Thus

$$\text{MR } (a, b, c, d, e) - \text{MR } (e, a, b) < \text{MR } (a, b, c, d) - \text{MR } (a, b).$$

In other words

$$\text{MR } (c, d/e, a, b) < \text{MR } (c, d/a, b).$$

Since MR $(c, d/a, b, e) = 1$, MR $(c, d/a, b) = 2$. Thus (a, b) and (c, d) are independent generics of D^2.

Thus C is an ample family of curves. We have now proved the following lemma.

Lemma 3.2 *Suppose D is a Zariski geometry. Then D is non-locally modular if and only if it is ample.*

We can now state the fundamental theorems of Hrushovski and Zilber.

Theorem 3.3 [HrZi 96] *1) If D is a non-locally modular Zariski geometry, then D interprets an algebraically closed field K. Moreover, if there is a very ample family of curves on D^2, then there is a smooth (quasi-projective) algebraic curve C over K and a bijection $f : D \to C$ such that, for all n, the bijection from D^n to C^n induced by f is closed and continuous.*

2) If K is a field interpretable in the Zariski geometry D, then K is a pure field. (i.e. any subset of K^n definable in D is already definable in the field language).

If we only know that D is non-locally modular (or, alternatively, that there is an ample family of curves on D^2), then we can find a map $f : D^n \to \mathbb{P}^n_K$ which is continuous and takes constructible sets to (algebraically) constructible sets.

4 Brief remarks on the proofs of Theorem 3.3

Specializations are a key tool in the proof of Theorem 3.3.

Let M be an elementary extension of D. If $A \subset M$, we say that $f : A \to M$ is a *specialization* if for any $a_1, \ldots, a_n \in A$ and any $\emptyset$-closed set $C \subseteq M^n$, if $\overline{a} \in C$, then $f(\overline{a}) \in C$.

We will show how in one basic situation specializations can be used to define tangency. Let Q_d be the family of all degree d curves in $\mathbb{P}^2$ that pass through 0. By identifying curves with their defining equations, Q_d becomes a projective variety.

Lemma 4.1 *Let $C \in Q_d$ be generic, let p be a generic point of C, and let l be the line through p and 0. Suppose $C' \in Q_d$ and l' is a line through 0 such that*

$$(C, p, l) \to (C', 0, l')$$

is a specialization. Then l' intersects C' at 0 with multiplicity at least two. In particular, if 0 is a simple point of C', then l' is tangent to C' at 0.

Proof We consider only the affine part of C. Suppose C is given by the equation $F(X, Y) = 0$ where

$$F(X, Y) = \sum_{i+j \leq d} a_{i,j} X^i Y^j.$$

Let $\overline{a} = (a_{i,j} : i + j \leq d)$. Let l be the line $Y = mX$, and let $f(T) = F(T, mT)$. There are polynomials $q_0, \ldots, q_{d-1}$ such that

$$f(T) = T \sum_{k=0}^{d-1} q_k(\overline{a}, m) T^k.$$

If $p = (x, y)$, then $f(x) = 0$ so $T - x$ divides $f(T)/T$. There is a polynomial $r(\overline{Z}, U, V)$ such that $T - u$ divides $\sum_{k=0}^{d-1} q_k(\overline{z}, v) T^k$ if and only if $r(\overline{z}, u, v) = 0$; r is the resultant of $T - U$ and $\sum_{k=0}^{d-1} q_k(\overline{Z}, V) T^k$ (see [Gr] II.2.2). Thus $r(\overline{a}, x, m) = 0$.

Let C' be given by

$$F'(X, Y) = \sum_{i+j \leq d} a'_{i,j} X^i Y^j$$

and let l' be $X = m'Y$ (a separate argument is needed for $Y = 0$). Since we have a specialization, $r(\overline{a}'_{i,j}, 0, m) = 0$. Thus T^2 divides $F(T, mT)$, and l' intersects C' at $(0, 0)$ with multiplicity at least 2.

Corollary 4.2 *Suppose* 0 *is a simple point of* C'_0 *and* $C'_1 \in Q_d$*. Let* C_0 *and* $C_1 \in Q_d$ *be independent generics. Suppose* $p \in C_0 \cap C_1$ *is such that* $p \neq 0$*. If*

$$(C_0, C_1, p) \to (C'_0, C'_1, 0)$$

is a specialization, then C'_0 *and* C'_1 *are tangent at* 0.

Proof

In a complete Zariski geometry any partial specialization can be extended (this is an easy exercise). Let l be the line through p and 0. There is a line l' through 0 such that

$$(C_0, C_1, p, l) \to (C'_0, C'_1, 0, l')$$

is a specialization. By 4.1, each C'_i is tangent to l' at $(0, 0)$.

The proof of theorem 3.3 1) is in the spirit of the proofs of the main results from [Rab] and (in a simpler setting) [MaPi]. In each case we try to detect the presence of a field by finding the forking pattern of Hrushovski's group (or field) configuration (see [Bous2 89]).

In [HrZi 96] the authors give a slightly different approach.

Consider the matrix $(a_{i,j} : i \in I, j \in J)$. We say that the matrix is *rank indiscernible* of type $p(m, n)$ if whenever $I_0 \subseteq I$, $J_0 \subseteq J$, $|I_0| = n$ and $|J_0| = m$, then MR $(a_{i,j} : i \in I_0, j \in J_0) = p(m, n)$.

We say that the matrix is an *indiscernible array* if for every $I_0 \subseteq I$ and $J_0 \subseteq J$ if $|I_0| = n$ and $|J_0| = m$, the type of $(a_{i,j} : i \in I_0, j \in J_0)$ depends only on n and m.

Lemma 4.3 [HrZi 96] *Assume we are working in D^{eq} for D a strongly minimal set.*

i) Suppose $(e_{i,j} : 1 \leq i \leq 4, 1 \leq j \leq 3)$ is rank indiscernible of type $m+n-1$. Then there is a definable rank one group.

ii) Suppose there is an infinite indiscernible array which is rank indiscernible of type $2m + n - 2$. Then there is an interpretable field.

Let p be a generic point in D^2. If there is an ample family of curves in D^2 then we can find a non-trivial rank 1 family of curves through p. If C_0 and C_1 are two such curves, we can define the composition

$$C_1^{-1} \circ C_0 = \{(x, y) : \exists z \ (x, z) \in C_0 \wedge (y, z) \in C_1\}.$$

This is a (possibly reducible) curve through (a, a). Hrushovski and Zilber define a notion of tangency, in a manner similar to the one above, so that they can find a family of curves $C(e_{i,j})$ such that $C^{-1}(e_{i,j}) \circ C(e_{i',j})$ and $C^{-1}(e_{i,j'}) \circ C(e_{i',j'})$ are tangent if and only if $i < i'$ and $j < j'$. This gives rise to an array as in 4.3 i) and allows us to construct an interpretable strongly minimal abelian group.

Using addition and composition we find a more complicated rank indiscernible array satisfying 4.3 ii) this gives rise to a definable field.

The proof of Theorem 3.3 2) necessitates developing an intersection theory for curves in D^2 paralleling the classical theory. Using a version of Bezout's theorem one finds that there is no room for any non-algebraic curves.

I refer the reader to [HrZi 93] for a more detailed outline of these proofs.

5 Generalizations

In [Hr 96] Hrushovski must work with a more general notion of Zariski geometry. Suppose T is a stable theory with quantifier elimination. Let M be a very saturated model of T. Recall that a complete type p is said to be minimal if for any definable subset X of M, $p(M) \cap X$ is finite or cofinite. If p is minimal, if D is any definable subset of M^n, then the U-rank of $D' = p(M)^n \cap D$, denoted $\mathrm{RU}(D')$, is finite (see for example [De], section 5).

Let p be a minimal type and let P be the realizations of p in some very saturated model. We say that a subset of P^n is closed if it is defined by a positive quantifier free formula (i.e. it is the trace on P^n of a set defined by a positive quantifier free formula). We say that p is *Zariski* if the following conditions hold.

i) Every closed set in P^n is a finite union of irreducible closed sets.

ii) If $X \subset Y$ are closed subsets of P^n and Y is irreducible, then $\mathrm{RU}(X) < \mathrm{RU}(Y)$.

iii) If X is a closed irreducible subset of P^n, $\mathrm{RU}(X) = m$ and $Y = \{\overline{x} \in P^n : x_i = x_j\}$, then $\mathrm{RU}(W) \geq m - 1$ for every non-empty irreducible component W of $X \cap Y$.

By ii) the family of closed sets determine a Noetherian topology. We have the following remnant of quantifier elimination: if $Y \subset P^n$ is definable then there is F a proper closed subset of the closure $\overline{Y}$ of Y such that $Y \setminus F = \overline{Y} \setminus F$. In [De] we will see that Zariski types arise in separably closed fields.

Note that even if p is Zariski, the structure $\mathcal{P}$ with universe P and relation symbols for all closed subsets of P^n may not be a Zariski geometry (as defined in §1). For example, let K be a non-algebraically closed, separably closed field of characteristic q and finite degree of imperfection, let $k = \bigcap K^{q^n}$ and let p be the type of a transcendental element of k. We will see in [De], that p is Zariski. Let $X = \{(x, y, z) \in P^3 : xz = y\}$. Then $\pi(X) = \{(x, y) : \exists z \in P\ xz = y\}$ is P^2 with infinitely many lines deleted. The closure of $\pi(X)$ is P^2 but there is no closed $W \subset P^2$ with $P^2 = W \cup \pi(X)$.

The following result is the analog of Theorem 3.3 for Zariski types.

Theorem 5.1 [Hr 96] *Let T be a stable theory, let p be a minimal type. Suppose p is Zariski and non-locally modular. Then T interprets a field F, with definable subfields F_α such that $\bigcap_\alpha F_\alpha$ is minimal and non-orthogonal to p.*

In [HrZi 96] Theorem 3.3 is deduced from certain axiomatic properties of specializations. These properties can be deduced for Zariski types and Theorem 5.1 follows.

6 Two examples

In this section we will give two further examples of Zariski geometries. First, we will show that Zariski geometries arise naturally in differentially closed fields. We will then give an example of an ample but not very ample Zariski geometry.

Let K be a differentially closed field of characteristic zero. We refer to [Wo] and [Mar 96] for background on differentially closed fields. Let $D \subset K^n$ be a strongly minimal set. We say that $X \subseteq D^n$ is δ-closed if there are differential polynomials $p_1, \ldots, p_n \in K\{X_1, \ldots, X_n\}$ such that

$$X = \{\overline{x} \in K^n : \bigwedge p_i(\overline{x}) = 0\}.$$

As we remarked in §1, this is a Noetherian topology. Using quantifier elimination, it is easy to see that (Z0), (Z1), and (Z2) hold. The next theorem of Hrushovski and Sokolovic shows we can make (Z3) hold as well.

Theorem 6.1 *Let $D \subset K^n$ be strongly minimal. There is $F \subset D$ finite such that $D \setminus F$ satisfies (Z3) and hence is a Zariski geometry.*

The next corollary follows easily from Theorems 6.1, 3.3 and the fact that any finite rank field interpretable in a differentially closed field is definably isomorphic to the constants (see [Wo] or [Pi1 96]).

Corollary 6.2 *If K is a differentially closed field of characteristic zero and $D \subset K^n$ is a non-locally modular strongly minimal set, then D interprets an algebraically closed field F and F is definably isomorphic to the field of constants of K.*

We now outline the proof of Theorem 6.1. Let $D \subset K^n$ be strongly minimal. Let V_0 be the δ-closure of D. Using quantifier elimination, strong minimality, and perhaps throwing away finitely many points, we may assume that V_0 is irreducible and that D is an open subset of V_0. Say

$$D = \{x \in V_0 : \bigvee_{i=1}^{M} g_i(x) \neq 0\}.$$

Suppose g_1 is not identically zero on V. Let $S_0 = \{x \in V_0 : g_1(x) \neq 0\}$. Then $D \setminus S_0$ is finite. Let $V_1 = \{(x, y) : x \in V_0 \wedge g_1(x)y = 1\}$. Then V_1 is δ-closed and isomorphic to S_0.

Let $O_1 = K\{\overline{X}\}/I^\delta(V_1)$ be the differential coordinate ring of V_1. Since V_1 is strongly minimal, the transcendence degree of O_1 is finite; otherwise, some element of the ring is a generic of K and the Morley rank of V_1 is infinite.

Lemma 6.3 *Let R be a differential domain. Let $R\{x\}$ be the domain generated by adding an element x, which is differentially algebraic over R. Then there is $p \in R\{x\}$ and $n \in \omega$ such that*

$$R\{x\} \subseteq R[x, x', \ldots, x^{(n)}, \frac{1}{p}].$$

Proof Let n be least such that $x, x', \ldots, x^{(n)}$ are algebraic over R. Let m be least such that there are polynomials $q_0, \ldots, q_m \in R[x, \ldots, x^{(n-1)}]$ such that

$$\sum_{i=1}^{m} q_i {x^{(n)}}^i = 0.$$

Taking the derivative, we obtain

$$\sum_{i=1}^{m} (\delta(q_i){x^{(n)}}^i + i q_i {x^{(n)}}^{i-1} x^{(n+1)}) = 0.$$

Let $p(x) = \sum i q_i {x^{(n)}}^{i-1}$. By the minimality of m, $p(x) \neq 0$. We have

$$x^{(n+1)} = \frac{-\sum_{i=1}^{m} \delta(q_i)(x^{(n)})^i}{p(x)}$$

and $\sum_{i=1}^{m} \delta(q_i){x^{(n)}}^i \in R[x, \ldots, x^{(n)}]$. Now an easy induction using the quotient rule shows that for all $j > n$, $x^{(j)} \in R[x, x', \ldots, x^{(n)}, 1/p]$. It is also easy to see that $R[x, x', \ldots, x^{(n)}, 1/p]$ is closed under differentiation (again by the quotient rule).

A routine induction now yields the following corollary (Note: adding $1/p$ and $1/q$ is the same as adding $1/pq$.)

Corollary 6.4 *Suppose $K\{x_1,\ldots,x_n\}$ has finite transcendence degree over K. There are $p(\overline{x}) \in K\{x_1,\ldots,x_n\}$ and m such that*

$$K\{x_1,\ldots,x_n\} \subseteq K[x_1,x_1',\ldots x_1^{(m)},x_2,x_2',\ldots x_2^{(m)},\ldots,x_n,x_n',\ldots x_n^{(m)},\frac{1}{p(\overline{x})}].$$

Thus there is $p \in O_1$ such that $O_1[1/p]$ is a finitely generated K-algebra.

We next use the following standard fact from algebraic geometry. (See the appendix).

Corollary 6.5 *Let R be a finitely generated K-algebra. Then there is $q \in R$ such that $R[1/q]$ is a regular ring.*

In particular, the affine variety corresponding to $R[1/q]$ has no singular points. (We don't really eliminate the singularity, we just push it off to infinity.)

Thus there is $q \in O_1$ such that $O = O_1[1/q]$ is regular. Let $V = \{(\overline{x},y) : \overline{x} \in V_1, q(\overline{x})y = 1\}$. Then V is isomorphic to a cofinite open subset of V_1. By choice of O, there is a smooth algebraic variety W with coordinate ring isomorphic to O. (Of course W will be a variety of much larger dimension.) If O^* is the differential coordinate ring of V^m, then O^* is the tensor product of O with itself m-times. Thus it is also the coordinate ring of W^m.

Let $C_0, C_1 \subseteq V^m$ be irreducible and δ-closed. Let $I_j = I^\delta(C_j)$. By the differential nullstellensatz ([Mar 96] 2.6), $I(C_0 \cap C_1) = \sqrt{I_0 + I_1}$. Let $P_1,\ldots,P_s$ be the primary decomposition of $\sqrt{I_0 + I_1}$. The P_i are differential prime ideals ([Mar 96] 1.19) . Let $F_i = V(P_i)$. Then the F_i are the irreducible components of $C_0 \cap C_1$.

We can now recast this as a geometric problem in the smooth variety W^m, or rather in the regular ring O^*. The dimension of an irreducible closed $C \subseteq W$ is equal to the Krull dimension of the corresponding ideal $I(C)$. By Lemma 1.2, we know that the dimension theorem holds for Zariski closed subsets of W.

Since $P_1 \cap \ldots \cap P_s$ is a decomposition of $\sqrt{I_0 + I_1}$ into prime ideals we must have

$$dim_{\mathrm{Krull}}\, P_1 \geq dim_{\mathrm{Krull}}\, I_0 + dim_{\mathrm{Krull}}\, I_1 - dim_{\mathrm{Krull}}\, \{0\}$$

where $dim_{\mathrm{Krull}}\, I$ is the Krull dimension of the ideal $I \subset O^*$.

The theorem now follows from the following lemma.

Lemma 6.6 *Let V be a strongly minimal, δ-closed set such that O, the differential coordinate ring of V, is a regular Noetherian K-algebra. Suppose that O has transcendence degree e over K. If $C \subseteq V^n$ is closed and $I = I^\delta(C)$, then*

$$dim_{\mathrm{Krull}}\, I = e \dim C = e \,\mathrm{MR}\, C.$$

Proof We first argue that $dim_{\mathrm{Krull}}\, I = e\,\mathrm{MR}\, C$. Let $(a_1,\ldots,a_m)$ be a generic point of C. Then $dim_{\mathrm{Krull}}\, I$ is equal to the transcendence degree of $K\langle\overline{a}\rangle$ over K (see 7.3).

Let $k = \mathrm{MR}\,(C)$. We may assume that $\mathrm{MR}\,(a_1,\ldots,a_k) = k$ and $a_{k+1},\ldots,a_m$ are model theoretically algebraic over $a_1,\ldots,a_k$. Since x is model theoretically algebraic over $K(\overline{y})$ if and only if x is field theoretically algebraic over $K\langle\overline{y}\rangle$, it follows that $K\langle a_1,\ldots,a_n\rangle$ is an algebraic extension (in the usual field theoretic sense) of $K\langle a_1,\ldots,a_k\rangle$. Since $a_1,\ldots,a_k$ are model theoretically independent and each $K\langle a_i\rangle$ has transcendence degree e over K, $K\langle\overline{a}\rangle$ has transcendence degree ek over K.

If $C_0 \subset C_1$ are irreducible, δ-closed sets, then

$$dim_{Krull}\ I^\delta(C_0) < dim_{Krull}\ I^\delta(C_1).$$

By the first claim, MR $C_0 <$ MR C_1. It easily follows that $dim\ C \leq$ MR C.

Finally we argue that MR $C \leq dim\ C$. This is really a more general phenomenon. Suppose $p \in S_n(K)$. Let

$$I_p = \{f \in K\langle\overline{X}\rangle : \text{“}f(\overline{v}) = 0\text{”} \in p\}.$$

Then MR $p \leq dim_{Krull}\ I_p$. This is proven by induction on α. Suppose MR $p = \alpha$. Let $I_p = \langle f_1,\ldots,f_M\rangle$. Let $\beta < \alpha$. By induction, there is a q with $p \neq q$ such that

$$\text{“}\bigwedge f_i(\overline{v}) = 0\text{”} \in q$$

and MR $q \geq \beta$; otherwise, this formula isolates p from other types of rank at least β and MR $p = \beta$. But then $I_q \supset I_p$ and, by induction, $dim\ I_q \geq \beta$. Thus, $dim\ I_p \geq \beta + 1$. So $dim\ I_p \geq \alpha$.

Finally, we give an example of an ample, non-very ample Zariski geometry. Let D be a Zariski geometry and let G be a group acting on D such that for each $g \in G$, the graph of $x \mapsto gx$ is closed and irreducible (i.e. $x \mapsto gx$ is a morphism). We say that the action is *semi-free* if whenever $gx = x$, then either $g = 1$ or $hx = x$ for all $h \in G$. Let $\tau : G^* \to G$ be a homomorphism with finite kernel H.

Let $D_0 = \{x \in D : \forall g \in G\ gx = x\}$. We assume D_0 is finite: otherwise, $D_0 = D$ and the action is trivial. Let D^* be a set on which G^* acts semifreely such that there are $|D_0|$ trivial orbits and $|D|$ regular orbits. Define E on D^* by xEy iff $x \in Hy$. Then G acts on D^*/E and the action is isomorphic to the action of G on D. Thus we can find $\sigma : D^* \to D$ surjective such that for $g \in G^*$, $\sigma gx = \tau(g)\sigma x$.

Topologize D^* by taking the weakest topology such that:
i) if $X \subset D^n$ is closed, then $\sigma^{-1}(X)$ is closed, and
ii) if $g \in G^*$, then $\{(x,y) : y = gx\}$ is closed.

Lemma 6.7 [HrZi 96] *D^* is a Zariski geometry.*

Let K be an algebraically closed field of characteristic zero and infinite transcendence degree. Let E be an elliptic curve defined over K with transcendental j-invariant (see [Ha] 317). Let a_1 and a_2 be independent generic points of E, let t_i be the translation $x \mapsto x + a_i$, and let G be the group generated by t_1 and

t_2. G is the free abelian group on two generators. Let G^* be the formal group with generators T_1 and T_2 and relations:

$$[T_1, T_2]^2 = [T_2, [T_1, T_2]] = [T_2, [T_2, T_1]] = 1.$$

Let H be the subgroup generated by $[T_1, T_2]$. Then $|H| = 2$ and $T_i \to t_i$ determines a homomorphism $\tau : G^* \to G$ with kernel H. Let E^* be the Zariski geometry determined by this data.

Theorem 6.8 [HrZi 96] *E^* is not interpretable in any algebraically closed field.*

Proof Suppose E^* is interpretable in an algebraically closed field F. Since E is interpretable in E^* and K is interpretable in E, F and K must be definably isomorphic ([Pil] 4.13). Thus, generically, E^* is a (possibly reducible) curve over K. Since $\sigma : E^* \to E$ is two-to-one E^* has at most two irreducible one dimensional components. Each T_i must either permute or preserve components. Thus $[T_1, T_2]$ must preserve components, a contradiction since $[T_1, T_2]$ acts non-trivially on fibers. Let C be the Zariski closure of the dimension one component of E^*. Without loss of generality C is non-singular.

Since there is a map $C \to E$, C must have genus at least 1. On the other hand, each T_i extends to an automorphism of C. Thus Aut(C) is infinite. This implies that C has genus at most one ([Ha] pg 348). Thus C is an elliptic curve.

Since C is isogenous to E, C has transcendental j-invariant. In particular $j \neq 0, 1728$. Thus Aut$(C) = \{x \mapsto \pm x + b : b \in C\}$ ([Ha] pg 321). If $T_i(x) = -x + b$, then $T_i^2(x) = -(-x + b) + b = x$, so we must have $T_i(x) = x + c_i$ for some c_i. But then T_1 and T_2 commute, a contradiction.

7 Appendix: classical dimension theory

We review some of the classical dimension theory of algebraic geometry. The following theorem is the fundamental result. We work over an algebraically closed field K.

Theorem 7.1 *Let V be a smooth variety of dimension m. Let $C_0, C_1 \subseteq V$ be irreducible varieties and let W be a nonempty irreducible component of $C_0 \cap C_1$. Then*

$$\dim W \geq \dim C_0 + \dim C_1 - m.$$

We begin by looking at three examples.

Example 1: $V = K^n$.

Consider $C_0 \times C_1 \subseteq K^{2n}$. Let $\Delta = \{(\overline{x}, \overline{y}) \in K^{2n} : \overline{x} = \overline{y}\}$. Then $C_0 \cap C_1$ is isomorphic to $(C_0 \times C_1) \cap \Delta$.

Let $W^* = \{(\overline{x}, \overline{y}) : \overline{x} = \overline{y} \in W\}$.

For each $i \leq n$, let W_i be the irreducible component of

$$(C_0 \times C_1) \cap \{(\overline{x}, \overline{y}) : \bigwedge_{j \leq i} x_j = y_j\}$$

which contains W^*. Then $W_n = W^*$. By induction, the next lemma implies that

$$dim\ W_i \geq dim\ C_0 + dim\ C_1 - i,$$

for all $i \leq n$.

Theorem 7.2 *Let $V \subseteq K^n$ be an irreducible variety. Let $H \subset K^n$ be a hypersurface. If W is an irreducible component of $V \cap H$, then* $dim\ W \geq dim\ V - 1$.

Proof See ([Sha] 1.6 Theorem 5)

Note that Theorem 7.2 has no smoothness assumption.

Example 2: Let $V \subset K^4$ be defined by $xy = zw$. Then $(0,0,0,0)$ is a singular point of V. Let

$$C_0 = \{(x,y,z,w) \in V : x = z = 0\}$$

and

$$C_1 = \{(x,y,z,w) \in V : y = w = 0\}.$$

Then $dim\ C_0 = dim\ C_1$, but $dim\ C_0 \cap C_1 = 0$.

What is the difference between these examples? In example 1 we were able to get the diagonal of V with $n = dim\ V$ equations. Here (at least near (0,0,0,0)), the diagonal of V requires 4 equations. We will see that smoothness implies that the number of hypersurfaces we need to intersect is at most the dimension.

Example 3: V is a smooth variety in $\mathbb{C}^n$ with $dim\ V = m$. Assume that $0 \in C_0 \cap C_1$.

By the implicit function theorem (and permuting coordinates if necessary), we can find $U \subset \mathbb{C}^m$ an open neighborhood of 0, and an analytic functions $f : U \to \mathbb{C}^{n-m}$, such that $f(0) = 0$ and for $\bar{t} \in U$,

$$(\bar{t}, f(\bar{t})) \in V.$$

Then near 0, $C_0 \cap C_1$ is isomorphic to

$$\{(\bar{t}, f(\bar{t}), \bar{s}, f(\bar{s})) : (\bar{t}, f(\bar{t})) \in C_0, (\bar{s}, f(\bar{s})) \in C_1\} \cap \Delta.$$

But, on this set, Δ is determined by the m-equations $t_1 = s_1, \ldots, t_m = s_m$. Thus the dimension of the intersection is (locally) at least $dim\ C_0 + dim\ C_1 - m$.

We next review the algebraic approach to dimension theory. (See for example chapter 11 of [AtMc].)

Let R be a Noetherian domain. Let $P \subset R$ be a prime ideal. The *Krull dimension* of P is the largest n such that there is a chain of prime ideals $P = P_0 \subset P_1 \subset \ldots \subset P_n$. The Krull dimension of R is equal to the Krull dimension of the ideal $\{0\}$.

Let $V \subseteq K^n$ be an irreducible variety. Let $O_V = K[X_1, \ldots, X_n]/I(V)$, O_V is the coordinate ring of the variety.

Lemma 7.3 *The dimension of V is equal to the Krull dimension of O_V. It is also equal to the transcendence degree of O_V over K.*

Suppose $0 \in V$. Let $O_{V,0}$ be the ring of rational functions on V which are defined at 0. Then $O_{V,0}$ is a local ring with maximal ideal $M = \{f \in O_V : f(0) = 0\}$. Suppose $I(V)$ is generated by $f_1, \ldots, f_m$. Let

$$f_i = \sum a_{i,j} x_j + \textit{ higher degree terms} \ldots.$$

Let $J = (a_{i,j})$. The tangent space at 0 is the kernel of the linear map $\overline{x} \mapsto J\overline{x}$. In particular, the dimension of the tangent space is equal to $n-\text{rank}(J)$.

On the other hand, consider the K vector space M/M^2. M is generated by $x_1, \ldots, x_n$ and we must mod out by the $f_1/M^2, \ldots, f_m/M^2$. But these are exactly the equations

$$\sum a_{1,j} x_j = 0$$

$$\vdots$$

$$\sum a_{m,j} x_j = 0.$$

This clearly has dimension $n - rank(J)$.

The smooth points of V are exactly the points where the dimension of the tangent space is equal to the dimension of V. This motivates the following definition.

Let R be a Noetherian local domain with maximal ideal M. We say that R is a *regular local ring* if the dimension of R is equal to the dimension of M/M^2. Then V is smooth at x iff $O_{V,x}$ (the localization of O_V at M_x the maximal ideal at x) is a regular local ring.

If R is a Noetherian domain we say that R is *regular* if and only if the localization at every maximal ideal is regular.

If x is a smooth point of V we say that $t_1, \ldots, t_m$ are a set of *local parameters* at x if they are a basis for M_x/M_x^2.

Lemma 7.4 *Suppose x is a simple point of V. Let $m = \dim V$ and let $t_1, \ldots, t_m$ be local parameters at x. Then there is an embedding of $O_{V,x}$ into $K[[t_1, \ldots, t_m]]$, the ring of formal power series in $t_1, \ldots, t_m$.*

Proof: See [Sha] II §2.2.

Lemma 7.4 allows us to extend the ideas from example 3 to a proof of Theorem 7.1. Let $t_1, \ldots, t_m$ be local parameters for $V \times V$ at (x, x). Then, there are $f_1, \ldots, f_n \in K[[\overline{t}]]$ such that $(f_1(\overline{t}), \ldots, f_n(\overline{t}))$ is a generic point of V. Then near (x, x) a point of $V \times V$ is of the form $(\overline{f}(\overline{t}), \overline{f}(\overline{s}))$ where $\overline{t}$ and $\overline{s}$ are local parameters. In this case the diagonal Δ is determined by the m-equations

$t_1 = s_1, \ldots, t_m = s_m$. Thus using 7.2 we see that each irreducible component of $C_0 \cap C_1$ has dimension at least $dim\ C_0 + dim\ C_1 - m$.

The following lemma will be useful.

Lemma 7.5 *Suppose R is a finitely generated K-algebra (i.e. R is the coordinate ring of an affine variety V). There is $a \in R$ such that $R[1/a]$ is a regular ring.*

Proof Let $R = K[X_1, \ldots, X_n]/I(V)$. The $f_1, \ldots, f_s$ generate $I(V)$. Let $m = dim\ V$. Let $g_1, \ldots, g_N$ be the determinants of $n - m \times n - m$ minors of the Jacobian matrix. Then x is a simple point if and only if some $g_i(x) \neq 0$. Suppose at some simple point $g_1(x) \neq 0$. Let $V^* = \{(\overline{x}, y) : \overline{x} \in V, yg_1(\overline{x}) = 1\}$. Then V^* is an m-dimensional variety isomorphic to an open subset of V. It is easy to see that at each point the Jacobian matrix has rank $n - m$. The coordinate ring of V^* is $R[1/g_1]$ and this is a regular ring.

Remark The proof of 7.5 uses the following simple idea. Suppose V is an affine variety and U is the open subset of V where $g(\overline{x}) \neq 0$. Then there is an affine variety $V^* = \{(\overline{x}, y) : \overline{x} \in V \text{ and } yg(\overline{x}) = 1\}$. Then V^* is an affine variety isomorphic to U. Note that we can choose U such that V^* is smooth (We haven't really removed the singularities, we have just "pushed" them off to infinity).

References

[AtMc] M. F. Atiyah and I. G. Macdonald, *Introduction to Commutative Algebra*, Addison-Wesley, 1969.

[Bous2 89] E. Bouscaren, *The group configuration–after E. Hrushovski*, in The Model Theory of Groups, A. Nesin and A. Pillay ed., Notre Dame University Press, 1989.

[Bous] E. Bouscaren, *Proof of the Mordell-Lang conjecture for function fields*, this volume.

[Da] V.I. Danilov, *Algebraic Varieties and Schemes*, in Algebraic Geometry I, I.R. Shafarevich ed., EMS 23, Springer, 1994.

[De] F. Delon, *Separably closed fields*, this volume.

[Gr] P. Griffiths, *Introduction to Algebraic Curves*, Translations of Math. Mon. 76, AMS (1989).

[Ha] R. Hartshorne, *Algebraic Geometry*, Springer, 1977.

[Hr 96] E. Hrushovski, *The Mordell-Lang conjecture for function fields*, J.AMS 9 (1996), 667-690.

[HrSo] E. Hrushovski and Z. Sokolovic, *Strongly minimal sets in differentially closed fields*, preprint.

[HrZi 93] E. Hrushovski and B. Zilber, *Zariski Geometries*, Bulletin AMS 28 (1993), 315-323.

[HrZi 96] E. Hrushovski and B. Zilber, *Zariski Geometries*, Journal AMS 9 (1996), 1-56.

[Mar 96] D. Marker, *Model Theory of Differential Fields*, Model Theory of Fields, Lecture Notes in Logic 5, Springer, 1996.

[MMP] D. Marker, M. Messmer and A. Pillay, *Model Theory of Fields*, Lecture Notes in Logic 5, Springer, 1996.

[MaPi] D. Marker and A. Pillay, *Reducts of* $(\mathbb{C},+,\cdot)$ *which contain* $+$, J. Symbolic Logic 55 (1990), 1243-1251.

[Pi1 96] A. Pillay, *Differential algebraic groups and the number of countable differentially closed fields*, Model Theory of Fields, Lecture Notes in Logic 5, Springer, 1996.

[Pi2 96] A. Pillay, *Geometrical Stability Theory*, Oxford University Press, 1996.

[Pil] A. Pillay, *Model theory of algebraically closed fields*, this volume.

[Rab] E. Rabinovich, *Interpreting a field in a sufficiently rich incidence system*, QMW Press (1993).

[Sha] I.R. Shafarevich, *Basic Algebraic Geometry*, Springer, 1977.

[Wo] C. Wood, *Differentially closed fields*, this volume.

[Zie] M. Ziegler, *Introduction to stability theory and Morley rank*, this volume.

[Zil] B. Zilber, *Lectures on Zariski type structures*, lecture notes (1992).

Differentially closed fields

CAROL WOOD

This article provides several definitions and results involving differential fields, with references for proofs but with few proofs. No new results appear here; rather the attempt has been made to display information in the setting in which it is needed in order to make the account in this volume self contained in terms of definitions, notation, and results. The reader seeking more details should refer to Marker's and Pillay's articles [Mar 96], [Pil 96].

1 Notation and basic facts

Throughout these notes on differential fields we will consider only rings and fields containing the rationals. In particular **all fields under consideration have characteristic 0**. In this section we list various definitions and basic results, often without proof or attribution. Further information can be found in several places; most of the results in this section can be found in [Bu 94], [Ka], [Ko], [Mar 96] and in [Po 85]. We warn the reader that there is no standard notation for many of the notions below.

Definition 1.1 *Let R be a ring.*
A **derivation** *on R is an additive map δ on R such that for all x and y in R, $\delta(xy) = x\delta(y) + y\delta(x)$.*

A **differential ring** *(=* **δ-ring***) is a ring R equipped with a derivation δ. A* **differential field** *is a differential ring which is also a field.*

A **differential ideal I** *(=* **δ-ideal***) of a differential ring R is an (algebraic) ideal which is closed under the derivation δ.*

Examples of differential fields abound. The field of meromorphic functions on a given region of the complex plane is a differential field, as is the field of germs of meromorphic functions at a point, with δ the usual derivative in each case. Although many results about differential fields hold in the more general setting of finitely many commuting derivations [McGr], we consider here only the case of a single derivation δ.

Another class of examples of differential rings are the differential polynomial rings, as follows. Given a differential ring R, the **differential polynomial ring $R\{y\}$** in one differential indeterminate y is the ring $R\{y\} = R[y, \delta y, \delta^2 y, \ldots]$, i.e., the (usual algebraic) polynomial ring over R in the infinite set of indeterminates $y, \delta y, \delta^2 y, \ldots$ By extending δ to all of $R\{y\}$ in the obvious manner, we get that

$R\{y\}$ is a differential ring. For nonzero $f \in R\{y\}$ we define **ord(f)**, the order of f, to be the largest n, if any, such that $\delta^n y$ appears in f with non-zero coefficient. For $a \in R$ we take $ord(a) = -1$. Associated with the differential polynomial f is its **separant** S_f, the partial derivative of f with respect to $\delta^n y$, $n = ord(f)$. Similarly we denote the differential polynomial ring in differential indeterminates $y_1 \dots, y_n$ by $R\{y_1, \dots, y_n\}$.

Let $R \subset S$ be differential rings. Let $a \in S$. We associate to a a differential ideal $\boldsymbol{I(a)}$ or $\boldsymbol{I_R(a)}$ of $R\{y\}$ consisting of those $f(y, \delta y, \dots, \delta^n y) \in R\{y\}$ such that $f(a, \delta a, \dots, \delta^n a) = 0$. If $I(a) \neq 0$ then we say that a is **differentially algebraic** over R. If $I(a) = 0$ then a is **differentially transcendental** over R. For several elements $a_1, \dots, a_n$ we also consider the corresponding differential ideal of those elements of $R\{y_1, \dots, y_n\}$ which vanish at $a_1 \dots, a_n$, and say that $a_1, \dots, a_n$ are **differentially algebraically dependent** over R if this differential ideal is nonzero, and **differentially algebraically independent** otherwise. We say that S is **differentially finitely generated** over R provided there are $a_1, \dots, a_n \in S$ such that S is the differential ring $R\{a_1, \dots, a_n\}$ generated by $a_1, \dots, a_n$, i.e., the smallest differential subring of S containing R together with the a_i's. In case R and S are fields, we denote by $\boldsymbol{R\langle a\rangle}$ the smallest differential field containing $R \cup \{a\}$.

As in ordinary algebraic geometry, prime and radical ideals play a special role among polynomial ideals.

Definition 1.2 *A* **prime** *differential ideal in a differential ring R is a differential ideal P which is also a prime ideal: if $ab \in P$ for some $a, b \in R$, then $a \in P$ or $b \in P$.*
A **radical** *differential ideal is a differential ideal Q such that if $a^n \in Q$ for some $a \in R$ and some $n > 0$, then $a \in Q$.*
We associate three ideals to a given subset A of R:
(A) *is the smallest ideal of R containing A*
$\{A\}$ *is the smallest differential ideal of R containing A, which is easily seen to be the same as $(A, \delta A, \delta^2 A, ...)$.*
$\sqrt{\{A\}}$ *is the smallest radical differential ideal of R containing A.*

A straight-forward computation shows that the (ring-theoretical) radical of a differential ideal is again a differential ideal, so that $\sqrt{\{A\}}$ is got by taking the radical of $\{A\}$, i.e., without taking any further closure under δ.
Another basic fact is that $\sqrt{\{A\}}\sqrt{\{B\}} \subseteq \sqrt{\{AB\}}$. From these one shows easily that for K a differential field (the case of interest here), any radical differential ideal in $K\{y_1, \dots, y_n\}$ is the intersection of prime differential ideals. An analogue of the Hilbert basis theorem holds in this setting for **radical ideals**, but not for all ideals. This analogue implies the descending chain condition for zero sets of differential polynomials (or equivalently the ACC for radical differential ideals), as needed in applying the Zariski geometry notions to differential fields.

Theorem 1.3 Finite Basis Theorem. *Let I be a radical ideal in the ring $K\{y_1, \dots, y_n\}$, where K is a differential field. Then I contains a finite subset*

A such that $I = \sqrt{\{A\}}$.

To see that the analogous result fails for differential ideals in general, consider the differential ideal generated by the set of squares of the form $\{(\delta^n y)^2 | n = 0, 1, 2, \ldots\}$.

Corollary 1.4 *Every proper ascending chain of radical differential ideals in* $K\{y_1, \ldots, y_n\}$ *is finite.*

Corollary 1.5 *Every radical differential ideal in* $K\{y_1, \ldots, y_n\}$ *is an intersection of finitely many prime differential ideals.*

To distinguish between sets definable in the language of rings (or fields) and those defined in the languages of differential rings/fields, we refer to the former as **definable** or, for emphasis, **field-definable** and the latter as **δ-definable.**

Definition 1.6 *Given a differential field K, the* **constant field** *of K is the field* $C_K = \{a \in K : \delta(a) = 0\}$.

Note that the constant field is a field, indeed it is a differential field, and is δ-definable. Moreover, if $b \in L$ for some $L \supset K$ and b is algebraic over C_K, then $b \in C_L$, i.e., b is also a constant.

Next we define a notion for differential fields which is similar to that of algebraically closed for (plain) fields.

Definition 1.7 *A differential field K is* **differentially closed** *provided for all* $f, g \in K\{y\}$ *such that* $ord(f) > ord(g)$*, K contains an element a such that* $f(a) = 0$ *and* $g(a) \neq 0$ *.*

Clearly the notion of differentially closed is expressible by a family of sentences in the language of differential fields, and any differentially closed field is algebraically closed. Since anything algebraic over the constants is again constant, it follows that the constant field of a differentially closed field is also algebraically closed. "Closed" is something of a misnomer, since over *any* differential field K one can add new solutions to a system as above, whenever f has positive order. For this reason Kolchin chose the name "constrainedly closed" rather than "differentially closed".

The details of adding new solutions to $f(y) = 0$ and $g(y) \neq 0$, where $ord(f) > ord(g)$ and $ord(g) > 0$, are as follows: take an irreducible factor f_0 of f such that $ord(f_0) = ord(f)$; write $f_0 = f_0(y, \delta y, \ldots, \delta^n y)$, adjoin n algebraically independent elements $a_0, \ldots, a_{n-1}$ to K; let $X = a_n$ be a solution to the algebraic equation $f_0(a_0, \ldots, a_{n-1}, X) = 0$, and assign $\delta^i(a_0) = a_i$ for $i = 1, \ldots, n$; this determines uniquely a differential field extension $K\{a_0, \ldots, a_n\}$ of K in which a_0 solves the given system.

Combining this fact with knowledge of the definable closure for fields, we note that for any differential field K and any subset A of K, the **δ-definable**

closure of A, i.e., the definable closure in the context of differential fields, is the differential field $Q\langle A\rangle$ generated by A.

Theorem 1.8 *The theory DCF of differentially closed fields is complete and admits elimination of quantifiers, hence is model complete.*

Remark: This theorem can be proved in a manner very similar to that given for ACF (algebraically closed fields) by Pillay [Pil].

Corollary 1.9 Differential Nullstellensatz. *Let S be a finite system of differential equations and inequations in n differential indeterminates $y_1, \ldots, y_n$ over a differentially closed field K. If S has a solution in some differential field L with $K \subset L$, then S has a solution in K.*

In the model-theoretic analysis of DCF we require the following easy consequence of the above theorem.

Corollary 1.10 *The constant subfield C_K of a differentially closed field K has* **no additional structure** *beyond that of algebraically closed field, i.e., for subsets of $(C_K)^n$, the notions of δ-definable over K and definable over C_K coincide. Thus C_K, viewed as a δ-definable subset of K, is strongly minimal.*

Proof: Let D be a δ-definable subset of K^n with parameters from K, such that $D \subseteq (C_K)^n$. Since DCF has quantifier elimination, the set D is a Boolean combination of sets defined by $f = 0$, where $f \in K\{x_1, \ldots, x_n\}$. Replace all $f(x_1, \ldots, x_n, \delta x_1, \ldots, \delta x_n, \ldots)$ by $f(x_1, \ldots, x_n, 0, \ldots, 0, \ldots)$ in the definition of D, thereby obtaining a formula in the language of rings which defines a subset D' of K^n. Note that D' intersects $(C_K)^n$ in D. Now C_K and K are models of the ω-stable theory ACF. Since D is the intersection of a definable subset of the larger algebraically closed field K with the smaller C_K, it follows from ω-stability [Zie, Cor. 3.7] that D must be field-definable with parameters from C_K, as required. □

We include below this definability observation in the form needed in [Bous].

Corollary 1.11 *Let H be a δ-definable group in K, $H \subset {C_K}^n$ and let g be a δ-definable map from H into K^m.*
1) There is an algebraic group G defined over C_K such that $H = G(C_K)$, the C_K-rational points of G.
2) Let D be a finite subset of K such that g is defined over D. Then $H = E_1 \cup \ldots \cup E_r$, where each E_i is a C_K-definable subset of ${C_K}^n$, and for each i, $g \restriction E_i$ is a rational function defined over some $D' \supset D$.

Proof: Part 1) is a direct application of 1.10 together with Weil's theorem (see [Pil, Prop. 4.12]) which says that over an algebraically closed field of characteristic 0, any definable group is an algebraic group.
For 2), note that for each h in H, $g(h) \in Q\langle D, h\rangle$. In fact, for some fixed m,

$g(h)$ lies in $Q(D, \delta(D), \ldots, \delta^m(D), h)$, since $\delta(h) = 0$ and only finitely many iterates of δ on D occur in the definition of g. Thus we have that g is defined in the language of fields over a finite set containing D. □

The following is an easy application of 1.10 to definable subgroups of K^n.

Corollary 1.12 *Let G be a δ-definable subgroup of the additive group K^n, where K is differentially closed. Then G is a vector space over C_K.*

Proof: Since $n \cdot x \in G$ for every $x \in G$ and every positive integer n, the set $S = \{s \in C_K | s \cdot G \subseteq G\}$ is an infinite definable subgroup of C_K. The (plain) algebraically closed field C_K is strongly minimal and hence thas no proper infinite definable subgroup. Thus $S = C_K$. □

We end our list of basic results with one final analogy with algebraically closed fields.

Theorem 1.13 *The theory DCF of differentially closed fields has elimination of imaginaries.*

The proof can be found in [Po 85]. Also, the second proof indicated by Pillay for ACF in this volume generalizes readily to this setting ([Pil, Lemma 2.6]). Pillay's first proof ([Pil, Fact 1.6]) is not directly applicable since we are not in a strongly minimal setting.

2 Types over differential fields

Let K be a model of DCF, and let $a \in L$, where $L \supset K$, L a differential field. We wish to consider the 1-types over K realized in L. Since DCF has quantifier elimination, it suffices to consider the differential ideals over K of elements of L. If a is differentially transcendental over K, i.e., if $I_K(a) = 0$, then $K\langle a\rangle$ is K-isomorphic to $qf(K\{y\})$, the quotient field of $K\{y\}$, under an isomorphism which sends a to y. In particular there is only one such simple differentially transcendental extension, up to K-isomorphism. If a is differentially algebraic over K, then there is a differential polynomial $f(y)$ of least order n and of least degree k in $\delta^n y$ such that $f(a) = 0$. We call such f a *minimum polynomial* for a over K. Clearly f is an irreducible polynomial in $K[y, \delta y, \ldots, \delta^n y]$. In general $\{f\} \neq I_K(a)$, but something close in spirit is true: $I_K(a) = \{g | S^i g \in \{f\}$ for some $i\}$ where $S = S_f$, the separant of f. In particular this is not sensitive to choice of minimum polynomial f for a, and $K\langle a\rangle$ is K-isomorphic to $qf(K\{y\}/I_K(a))$.

Note also that if f is any irreducible differential polynomial in $K\{y\}$, then there is an element a in some differential field extension L of K such that f is a minimum polynomial for a over K. Such an extension was constructed in the previous section. Thus the differentially algebraic simple extensions are determined by differential polynomials over K. This, together with the fact that there is only one differentially transcendental simple extension, gives a total of $card(K)$ 1-types over K. Therefore:

Theorem 2.1 *The theory DCF of differentially closed fields is ω-stable.*

A result of Shelah guarantees the existence and uniqueness of prime model extensions for an ω-stable theory. The prime model extension of a differential field is called its **differential closure**, or **δ-closure**.

Theorem 2.2 *Every differential field k has a differential closure K, which is unique up to k-isomorphism. K is differentially algebraic over k, and the constant field C_K is algebraic over C_k.*

In [Man] Manin moved with success between the settings of fields and differential fields, using the derivative to maneuver inside an abelian variety. This strategy was also employed by Buium [Bu 94] and by Hrushovski [HrSo],[Hr 96], as will be apparent in this volume. Indeed we shall have occasion to move between δ-fields and fields, between DCF and ACF (algebraically closed fields), and to compare the ranks of formulas and types in the two settings. In order to distinguish the DCF and ACF notions we denote by RM_{field} the Morley rank in the language of (plain) fields, i.e. in the ACF setting, reserving RM for the Morley rank in the DCF setting.

First we consider RM_{field}, as described in [Pil]. Recall that all (complete) 1-types over an algebraically closed field C have Morley rank 0 except for the unique 1-type of a transcendental element, which has $RM_{field} = 1$. In the case of n-types, rank corresponds to transcendence degree over C of any realization. Transcendence degree also determines the rank of incomplete types. In particular, the rank of an algebraic variety V is equal to the maximum, over all extensions of C, of the transcendence degrees over C of points of V. In other words. the dimension of V is the same as $RM_{field}(V)$.

Next we consider possible values for Morley rank RM of 1-types over differentially closed K. If a has minimum polynomial over K of order n, then n is the transcendence degree of the differential field extension generated by a over K, hence n is an upper bound for the number of times that $q = type(a/K)$ can fork, giving that $RM(q) \leq n$. In particular, if a is algebraic over K, then $a \in K$ and $RM(a) = 0$. Since all the differentially algebraic types have finite rank, the unique differentially transcendental 1-type has rank at most ω. In fact all finite ranks are achieved; for example, if a has minimum polynomial $\delta^n y = 0$ over K, one can check that $RM(tp(a/K)) = n$. From this it follows that the differentially transcendental 1-type has rank ω. Notice also that $RM(K) = \omega$, and $RM(C_K) = 1$.

Now look back into the field setting, and consider the 1-type q_0 of a transcendental element over the algebraically closed field $C_K = C$. There is a unique non-forking extension of the field type q_0 to a (complete) differential field type

q over K, namely q is the type of the differential transcendental over K. Thus q_0 is a rank 1 type in the field setting, and is a rank ω partial type in the δ-field setting. Similarly if for $k > 0$ one starts with an n-type of rank k over C, or more generally an algebraic variety in n variables with generic point of transcendence degree k, then the corresponding n-type, resp. δ-definable set, will have Morley rank $\omega \cdot k$.

3 Fields interpretable in differentially closed fields

The next two sections are drawn from [Pil 96], any mistakes due to the present author of course.

We recall the following fact about definable fields [Pil, 4.13].

Theorem 3.1 *Let C be an algebraically closed field and let C' be an infinite field definable in C. Then C' is definably isomorphic to C.*

Strongly minimal δ-definable subsets in differentially closed fields turn out to be Zariski geometries [HrSo]. This is proved in this volume in Marker's contribution [Mar, Section 6]. When we apply the Zariski geometries methodology to the differential fields setting in [Bous], we will naturally be led to consider the finite Morley rank fields δ-definable in differentially closed fields. Constant fields are rank 1 such fields. By ω-stable or finite Morley rank "field" we mean a field with possibly additional structure. We have already noted that any model of DCF is algebraically closed, as is its constant field. These can be viewed as special cases of the following general result of Macintyre [Pil, 5.2].

Theorem 3.2 *An infinite ω-stable field is algebraically closed.*

Definition 3.3 *Let K be differentially closed and let G be δ-definable in K. We say G is* **δ-connected** *provided G has no δ-definable subgroup of finite index > 1. A group G is* **field-connected** *if G has no field-definable subgroup of finite index > 1.*

Trivially, δ-connected implies field-connected. It is also true (see Appendix C of [Mar 96]) that a definable, field-connected group is δ-connected.

Lemma 3.4 *Let K be differentially closed. Let G be a δ-connected group of finite Morley rank which is δ-definable in K. Then G is δ-definably embeddable into a field-connected group H which is definable (i.e., field-definable) in K. Moreover, if G is centerless, then H can be chosen to be a subgroup of $GL(n, K)$.*

Proof: Let K_0 be a countable differentially closed subfield of K over which G is defined. Since G has finite Morley rank, it must be that the generic p of G over K_0 has finite transcendence degree over K_0. Let a realize p, where a lies in some appropriately saturated model. Since the transcendence degree is finite, one can find $a_1, \ldots, a_n$ such that $K_0\langle a\rangle = K_0(a_1, \ldots, a_n) = K_0(f(a))$ for

some δ-definable $f : G \to K^n$. If b is generic over $K_0\langle a \rangle$, then $a \cdot b$ is generic over both $K_0(f(a))$ and $K_0(f(b))$, and moreover $f(a \cdot b) \in K_0(f(a), f(b))$. Let $g(x_1, \ldots, x_n, y_1, \ldots, y_n)$ be a function definable over K_0 such that $f(a \cdot b) = g(f(a), f(b))$. Let q_{field} be the type of $f(a)$ in the language of fields (i.e., no δ).Then $f(a)$ and $f(b)$ are independent realizations of q_{field}, and $g(x, y)$ defines a group multiplication generically on realizations of q_{field}. Then (see remark in [Pil] right after 4.11) there is some field-connected group H with q_{field} as generic. The map $f : a \mapsto f(a)$ is defined on generics of G and extends to give us the required δ-definable embedding ϕ of G into the definable connected group H. By Weil's theorem [Pil, 4.12] we can assume that H is an algebraic group defined over K.
Now assume further that G is centerless, and that H has been chosen of least dimension (= least RM_{field} such that G δ-embeds in H). Let $Z(H)$ be the center of H; then $H/Z(H)$ is also field-connected and is also (isomorphic to) an algebraic group. Now $\phi(G) \cap Z(H) = 1$ since G is centerless, and so G δ-embeds into $H/Z(H)$. By minimality of dimension of H, it must be that $Z(H)$ is finite, hence $H/Z(H)$ is centerless. Thus we may assume that G is embedded in a *centerless* algebraic group H. Any connected centerless algebraic group H embeds, as an algebraic group, in $GL(n, K)$ for some n [Pil, 4.10]. Therefore we have a δ-embedding of G into $GL(n, K)$. □

Remark: The restriction to the case where G has finite Morley rank is unnecessary, but the infinite case, not needed in the present volume, requires a different proof. For the stronger result see [Pi 95].

Theorem 3.5 (Cassidy-Sokolovic) *An infinite field F of finite Morley rank which is δ-definable in a differentially closed field K is δ-definably isomorphic to the constant subfield C_K.*

Proof: Since DCF is ω-stable, an infinite field F which is δ-definable in a differentially closed field K is also ω-stable. Hence F is algebraically closed, by Macintyre's theorem. Note also that the group $G = PSL_2(F)$ is simple, and we can recover F definably from the group structure of G; indeed, G is a finite product of conjugates of the abelian subgroup H of upper triangular matrices in G. Furthermore, if F has finite Morley rank, then G has finite Morley rank as well (cf. also Section 2 of [Zie]).
Applying the previous lemma to G, we can choose ϕ a δ-embedding of G into $GL(n, K)$, for some n. By altering ϕ by a conjugation (by, e.g., the Lie-Kolchin theorem) we can assume that ϕ sends the commutative group H into U_n, the group of upper triangular matrices of $GL(n, K)$.
Let π be the projection of U_n to the group of diagonal matrices, and let $A = ker(\pi) \cap \phi(H)$.
Case 1. A nontrivial:
Then A is a commutative group of unipotent matrices. Any such group is definably isomorphic to a subgroup of the vector space K^{n^2}. This gives us that

A is a δ-definable subgroup of K^{n^2}, hence A is a vector space over $C_K = C$, by Theorem 1.12. Since F has finite Morley rank, so does A, hence A is a finite dimensional vector space over C.
Since G is simple, it is generated by conjugates of A. Now we apply Zilber's Indecomposability Theorem [Las, 5.2] to conclude that only finitely many of these conjugates are needed, hence that G is contained in a quotient of a finite product of finite dimensional vector spaces over C. By elimination of imaginaries, we get that G is δ-isomorphic to a group δ-definable in C. But δ-definable in C is the same as definable in C, by 1.10.
Now recovering F from G we get that F is δ-isomorphic to a field definable in C, hence by 3.1 to C itself.
Case 2. A trivial:
Then π gives an isomorphism between $\phi(H)$ and a subgroup B of the multiplicative group $(K^*)^m$, where $m = n^2$. If $B \cap C^m$ is infinite, then we can apply Zil'ber's theorem as in Case 1 and conclude that F is δ-isomorphic to C. If $B \cap C^m$ is finite, say of order k, then $kB \cap C^m = 1$ and the logarithmic derivative (mapping x to $\delta x/x$) gives a δ-definable isomorphism between kB and an infinite subgroup of K^m, and we are again ready to apply Zil'ber's Theorem as above. □

Note that by the results of Section 1, a field of finite Morley rank as in the theorem will be a (plain) algebraically closed field, with *no additional structure.*

4 The Manin kernel

Our goal in this section is to display the ingredients, both model-theoretic and geometric, while refering to [Pil 96] for details, of the following:

Theorem 4.1 *Let K be differentially closed. Let A be an abelian variety over K and let Γ be a subgroup of $A(K)$ of "finite rank". Then there is a δ-definable subgroup H of A such that $\Gamma \subset H$, and such that H has finite Morley rank.*

"Finite rank" is placed in quotes above, to distinguish it from the model theoretical notions of rank introduced earlier. Here we mean that Γ contains a finitely generated subgroup Γ_0, such that for each $x \in \Gamma$ there is some integer $n \geq 1$ such that $nx \in \Gamma_0$. The group H in the theorem is related to the kernel of a homomorphism introduced by Manin [Man] and usually called the Manin kernel. This figures in Buium's and Hrushovski's work on the characteristic zero function field case of the Mordell-Lang conjecture.

Definition 4.2 *Given $n \geq 0$ and $m > 0$, the maps $e_m : K^n \to K^{n(m+1)}$ are defined by*

$$e_m(\overline{a}) = e_m(a_1, \ldots, a_n) = (a_1, \ldots, a_n, \delta a_1, \ldots, \delta a_n, \ldots, \delta^m a_1, \ldots, \delta^m a_n).$$

Let G be a δ-definable connected group, $G \subset K^n$ for some n. Taking the maps as above we endow each $e_m(G)$ with a group structure by defining

$$e_m(\overline{a}) * e_m(\overline{b}) = (\overline{a} \cdot \overline{b}, \delta(\overline{a} \cdot \overline{b}), \ldots, \delta^m(\overline{a} \cdot \overline{b})).$$

The resulting group structure on $e_m(G)$ is δ-connected and δ-definable, and e_m is a δ-definable isomorphism from G onto $e_m(G)$. So why should we bother to find all these "new" groups, when each turns out to be isomorphic to the original G? What we intend is to extend the scope of $*$ beyond $e_m(G)$. Suppose from now on that G and its group structure are *definable*, not just δ-definable. A motivating example for this is the case where G is an abelian variety.

Having G's multiplication definable does not imply that $e_m(G)$ is definable, but it does imply that the **multiplication** $*$ on $e_m(G)$ is definable. Let p be the generic type of $e_m(G)$ and let p_{field} be the restriction of p to the field language. Then (see the remark in [Pil] right after 4.11 and 4.12) there is a field-connected definable group G_m (i.e., a connected algebraic group G_m) such that p_{field} is the generic of G_m and such that the group operation on G_m agrees with $*$ on independent realizations of p.
Since the group operation on G is definable it is possible to show that there is a definable embedding, first of independent generics, then of all of $e_m(G)$ into G_m, such that the group operation on G_m restricted to $e_m(G)$ is in fact $*$. By abuse of notation we identify $e_m(G)$ with a δ-definable subgroup of G_m.

Thus from our original G we have obtained a definable group G_m for each m, together with a δ-definable embedding of G into G_m. Next we define a map $\pi_m : G_m \to G_{m-1}, m > 0$, taking $G_0 = G$. We start with the map from $e_m(G)$ to $e_{m-1}(G)$ given by restriction to the first nm coordinates. This determines the map π_m generically, and it extends to a definable map from G_m onto G_{m-1}.

Fact 4.3 *For all* $m > 0, RM_{field}(G_m) = (m+1)RM_{field}(G)$.
For $X \subset G$ *such that* X *is* δ*-definable, there is some* m *such that there exists* $Y \subset G_m$ *with* Y *definable and* $e_m(X) = Y \cap e_m(G)$.
If X *is a* δ*-definable subgroup, then* Y *can be taken to be a definable subgroup.*
If X *is* δ*-connected, then* Y *can be taken to be field-connected.*

Definition 4.4 *Let* ρ_m *be the composition* $\pi_m \circ \pi_{m-1} \circ \ldots \circ \pi_1 : G_m \to G$.

An easy example of a group of rank r over a field is a vector group of dimension r, i.e., K^r with coordinatewise addition. We will see that the increase in Morley rank in going from G to G_m comes largely from just such considerations.

Fact 4.5 *There is no nontrivial definable homomorphism from a vector group into an abelian variety.*

We also require the following result of Rosenlicht [Ro 58] concerning dimension of extensions of an abelian variety by a vector group.

Fact 4.6 *Given an exact sequence of groups $0 \to B \to G \to A \to 0$, where A is an abelian variety, G is an algebraic group, and B is a vector group, there is a connected algebraic subgroup G^* of G such that G^* maps onto A and $RM_{field}(G^*) \leq 2 \cdot RM_{field}(A)$.*

One last ingredient is to know that the kernels of the ρ's are well behaved:

Fact 4.7 *If G is a commutative algebraic group of dimension n, i.e., such that $RM_{field}(G) = n$, then the kernel J_m of ρ_m is a vector group of dimension $m \cdot n$.*

Remark about the proof of 4.7: Since G is commutative, so must be G_m. It is easy to see that the kernel of each π_m is a vector group. From the resulting exact sequence of vector groups, one obtains that the kernel of ρ_m is also a vector group (see for example [Se, Chapter VII, section 2.7]).

Now we are ready to assemble the various pieces to prove the main theorem.

Lemma 4.8 *Let G be a connected commutative algebraic group, and let H be a δ-connected δ-definable subgroup, Zariski-dense in G. Then G/H is δ-definably isomorphic to a vector group J.*

Proof: Since G is definable, we have definable groups G_m and δ-definable maps e_m as above. By Fact 4.3 there exist an integer m and a group H^*, definable in G_m and field-connected, such that $e_m(H) = H^* \cap e_m(G)$.
Now let J_m be the vector group which is the kernel of ρ_m. Since H is dense in G, we get that $\rho_m(H^*) = G$ and that $G_m = J_m \cdot H^*$. Since J_m is a vector group, we can find a complement J (again obviously a vector group) of $J_m \cap H^*$ in J_m. Therefore G_m is the direct product of J and H^*. Let $\pi : G_m \to J$ be projection. The homomorphism $\phi : G \to J$ given by $\phi(a) = \pi(a, \delta a, \ldots, \delta^m a)$ has kernel H, as required. □

Armed with this information about δ-definable subgroups, we now proceed to show that these subgroups exist.

Proposition 4.9 *Let A be an abelian variety defined over K. Then there exists a δ-definable homomorphism ϕ from A into K^n for some n, such that the kernel of ϕ is an infinite group of finite Morley rank.*

Proof: Let A_m and ρ_m be as above, with $\rho_m : A_m \to A$. Suppose B and B' are definable subgroups of A_m which are each mapped onto A by ρ_m. Then $Ker(\rho_m)$ is a vector group into which A_m/B and A_m/B' embed, as in 4.8. We can combine these two to get an embedding of $A_m/(B \cap B')$ in $Ker(\rho_m)$. This implies that $B \cap B'$ also maps onto A, by Fact 4.5. (Otherwise we would get a nontrivial homomorphism of a vector group into A.) Thus there is a unique minimal definable subgroup B_m of A_m mapping onto A under ρ_m. By unicity, it must be that $\pi_m(B_m) = B_{m-1}$. Note also that B_m is infinite: since A_m/B_m embeds in a vector group, B_m must contain all torsion points of A_m. There are infinitely many torsion points in A, and B_m must contain the infinite set of the images of

those torsion points under e_m. Let $B = \{a \in A : (a, \delta a, \ldots, \delta^m a) \in B_m$ for all $m\}$. (There is a slight cheat here; we should be taking a generic such element, noting that we can define a group B which agrees on generics with the structure given on B_m for all m.) Note that B is the intersection of infinitely many δ-definable groups. Since DCF is ω-stable, we have that B is itself δ-definable, by [Las, 2.4]. Indeed, for sufficiently large m, $B = \{a \in A : (a, \delta a, \ldots, \delta^m a) \in B_m\}$. Fix m large enough, and let $\phi : A \to A_m/B_m \subset K^n$. Then ϕ is δ-definable and B is the kernel of ϕ.
It remains to see that B has finite Morley rank. Note that Fact 4.6 gives a uniform upper bound d in terms of A, namely $2 \cdot RM_{field}(A)$ for the dimension of the B_m's. Let k be a small subfield over which B is defined. For any $b \in B$, the field $k < b >$ has finite transcendence degree over k, indeed has transcendence degree bounded above by d. Thus $RM(b) \leq d$ for any $b \in B$, and so $RM(B) \leq d$. □

Pillay has noted that this upper bound can be improved, by a different argument, to $d/2$, i.e., to $RM_{field}(A)$.

We are now ready to complete the proof of Theorem 4.1: The image of the given Γ under the δ-definable map ϕ is a "finite rank" subset of the vector group, hence is contained in a subgroup X of K^n which has finite dimension over the field of rationals. The inverse image $H = \phi^{-1}X$ is therefore a δ-definable subgroup of A of finite Morley rank containing Γ, as desired. □

Corollary 4.10 *If A is a simple abelian variety, then A contains an infinite δ-connected δ-definable subgroup B_{min} of finite Morley rank which is minimal among all infinite δ-definable subgroups of A. Moreover, the group B_{min} contains the torsion points of A.*

Proof: As in the proof of 4.9, if G maps to a vector group and H is the kernel, then H contains all the torsion of G; if we take G to be an abelian variety, then the torsion part of G is infinite, hence H is infinite. Note also that any infinite δ-connected δ-definable subgroup B of a simple abelian variety A is Zariski dense in A. By 4.8, any such B must contain all the torsion of A.
By 4.9, there exist δ-connected δ-definable infinite subgroups of A of finite Morley rank. The intersection B_{min} of all δ-connected δ-definable subgroups of A is again infinite, since it contains the torsion of A. We note that B_{min} is δ-connected, and is contained in a δ-definable group of finite Morley rank. Furthermore, B_{min} is δ-definable, since we are working in an ω-stable setting [Las, 2.4]. As a δ-definable subgroup of a group of finite Morley rank, B_{min} has finite Morley rank. Clearly B_{min} is minimal as claimed. □

References

[Bous] E. Bouscaren, *Proof of the Mordell-Lang conjecture for function fields*, this volume.

[Bu 94] A. Buium, *Differential Algebra and Diophantine Geometry*, Hermann, 1994.

[HrSo] E. Hrushovski and Z. Sokolovic, *Strongly minimal sets in differentially closed fields*, preprint.

[Hr 96] E. Hrushovski, *The Mordell-Lang conjecture for function fields*, J. AMS 9 (1996), 667-690.

[Ka] I. Kaplansky, *An Introduction to Differential Algebra*, Hermann, 1957.

[Ko] E. Kolchin, *Differential Algebra and Algebraic Groups*, Academic Press, 1973.

[Las] D. Lascar, *ω-stable groups*, this volume.

[Man] Y. Manin, *Rational points of algebraic curves over function fields,* Isvetzia 27(1963), 1395-1440 (AMS Transl. Ser II 50(1966) 189-234).

[Mar 96] D. Marker, *Model Theory of Differential Fields*, Model Theory of Fields, Lecture Notes in Logic 5, Springer, 1996.

[Mar] D.Marker, *Zariski geometries*, this volume.

[McGr] T. McGrail, *Model Theory of Partial Differential Fields,* Ph.D. Thesis, Wesleyan University, 1997.

[Pi 95] A. Pillay, *Model theory, differential algebra, and number theory*, Proceedings ICM (Zurich,1994), Birkhauser, 1995.

[Pil 96] A. Pillay, *Differential algebraic groups and the number of countable differentially closed fields*, Model Theory of Fields, Lecture Notes in Logic 5, Springer, 1996.

[Pil] A. Pillay, *Model theory of algebraically closed fields*, this volume.

[Po 85] B. Poizat, *Cours de Théories des Modèles*, Nur alMantiq walMa'arifah, Villeurbanne, France, 1985.

[Ro 58] M. Rosenlicht, *Extensions of Vector Groups by Abelian Varieties*, American Journal of Math. 80 (1958), 685-714.

[Se] J.P. Serre, *Algebraic Groups and Class Fields*, Springer, 1988.

[Zie] M. Ziegler, *Introduction to Stability theory and Morley rank*, this volume.

Separably closed fields

FRANÇOISE DELON

Separably closed fields are stable. When they are not algebraically closed, they are rather complicated from a model theoretic point of view: they are not superstable, they admit no non trivial continuous rank and they have the dimensional order property. But they have a fairly good theory of types and independence, and interesting minimal types. Hrushovski used separably closed fields in his proof of the Mordell-Lang Conjecture for function fields in positive characteristic in the same way he used differentially closed fields in characteristic zero ([Hr 96], see [Bous] in this volume). In particular he proved that a certain class of minimal types, which he called thin, are Zariski geometries in the sense of [Mar] section 5. He then applied to these types the strong trichotomy theorem valid in Zariski geometries.

We will recall here the basic algebraic facts about fields of positive characteristic (section 1) and reprove classical model-theoretical results about separably closed fields. We will consider only the non perfect fields of finite degree of imperfection, which are the ones appearing in the proof of Hrushovski and which admit elimination of quantifiers and imaginaries in a simple natural language (section 2). We will then develop a general theory of "λ-closed subsets" and associated ideals (sections 3 and 4), which has the flavour of the classical correspondance between Zariski closed subsets and radical ideals in algebraic geometry, and which allows us to prove that all minimal types are Zariski (section 5). Finally, following Hrushovski, we define thin types and explain how algebraic groups give rise to such types (section 6).

Notation: we use the notation $\bigwedge$-definable to mean infinitely definable.

Many thanks to Gabriel Carlyle, Marcus Tressl, Carol Wood and especially Elisabeth Bouscaren and Zoé Chatzidakis for reading preliminary versions of this text and detecting many insufficiencies and/or redundancies.

1 Fields of characteristic $p > 0$

Except for facts 1.3 and 1.4, everything in this section is classical and can be found in [Bour] and [Lan1 65].

Each field $K \neq \mathbb{F}_p$ has a non trivial endomorphism, the Frobenius map $x \to x^p$. Hence $K^p := \{x^p; x \in K\}$ is a subfield of K. K is said to be <u>perfect</u> if $K^p = K$. For each $x \in K$ there is, in every algebraically closed field containing K, a unique y satisfying $y^{p^n} = x$. We will denote $y = x^{p^{-n}}$ and $K^{p^{-n}} := \{x^{p^{-n}}; x \in K\}$.

An irreducible polynomial over K can have multiple roots: for example $X^p - a$ is irreducible over K iff $a \in K \setminus K^p$. A polynomial is separable if all its roots are distinct. Let x be algebraic over K, f its minimal polynomial; x is said to be separable over K if f is separable, or equivalently if x has exactly degree(f) distinct conjugates over K; x is purely inseparable over K if $f = X^{p^n} - a$ for some integer $n \geq 1$ and some $a \in K \setminus K^p$, or equivalently if every conjugate of x is equal to x. In general f may be written as $f(X) = g(X^{p^n})$, where $g \in K[X]$ is separable, hence the extension $K \subseteq K(x)$ may be decomposed as

$$K \subseteq K(x^{p^n}) \subseteq K(x).$$

More generally, every algebraic extension $K \subseteq L$ may be decomposed as $K \subseteq L_1 \subseteq L$ where the extension $K \subseteq L_1$ is separable (:$\Leftrightarrow$ every $x \in L_1$ is separable over K) and $L_1 \subseteq L$ is purely inseparable (:$\Leftrightarrow$ every $x \in L$ is purely inseparable over L_1).

The set of separably algebraic elements over K form a subfield K^s of the algebraic closure K^a. Purely inseparable elements form a subfield $\bigcup_{n \in \mathbb{N}^*} K^{p^{-n}} =: K^{p^{-\infty}}$ of K^a. Clearly $K^a = K^s.K^{p^{-\infty}}$ and K^s and $K^{p^{-\infty}}$ are linearly disjoint over K.

K^s is called the separable closure of K, and K is said to be separably closed if $K = K^s$.

$K^{p^{-\infty}}$ is called the perfect closure of K.

Theorem Every finite separable extension of K is of the form $K[x]$.

This well known "primitive element theorem" does not hold in general for non separable extensions. For example, if $K = \mathbb{F}_p(X, Y)$ with X and Y algebraically independent over $\mathbb{F}_p$, and $L = K^{p^{-1}}$, one has $[L : K] = p^2$ but $[K[x] : K] = p$ for every $x \in L \setminus K$.

The relation of p-dependence, which we will define now, is adequate for describing this phenomenon.

Let $A, B \subseteq K$ and $x \in K$. We say x is p-independent over A in K if $x \notin K^p(A)$; B is p-free over A if b is p-independent over $A \cup (B \setminus \{b\})$ for all b in B. We say "p-independent" or "p-free" instead of p-independent or p-free over $\emptyset$. B p-generates K if $K \subseteq K^p(B)$. Now, in K, B is p-generating minimal iff it is p-free maximal iff it is p-free and p-generating. Such a B is called a p-basis of K. All p-bases of K have the same cardinality. If ν is this cardinality, ν is finite iff $[K : K^p]$ is finite, and in this case $p^\nu = [K : K^p]$. We call ν the degree of imperfection of K, where $\nu \in \mathbb{N} \cup \{\infty\}$.

$B = \{b_i; i \in I\}$ is a p-basis of K iff the monomials $m_j = m_j(B) := \prod_{i\in I} b_i^{j(i)}$, for j any map from I into $\{0,1,...,p-1\}$ with finite support, form a linear basis of the K^p-vector space K; B is a p-basis of K iff for any integer n, the set $\{m_j(B); j \in \{0,1,...,p^n-1\}^I$ with finite support$\}$ forms a linear basis of the K^{p^n}-vector space K. Consequently, if B is a p-basis of K, then any $x \in K$ can be written uniquely as $x = \sum x_j^p m_j$, with $j \in \{0,1,...,p-1\}^I$ having finite support, and with $x_j \in K$, almost all zero. The x_j's are called the components of x with respect to B, or its p-components.

We can now define and characterize separable extensions which are not necessarily algebraic.

An extension $K \subseteq L$ is called separable if one of the following equivalent conditions holds, where all fields below are subfields of L^a: (i) $K^{p^{-1}}$ and L are linearly disjoint over K
(ii) $K^{p^{-\infty}}$ and L are linearly disjoint over K
(iii) every p-free subset of K is p-free in L
(iv) some p-basis of K is p-free in L.

Remarks:
1. The two definitions of separability coincide for algebraic extensions.
2. Purely transcendental extensions are separable. More precisely, if B is a p-basis of K and X is algebraically free over K, then $B \cup X$ is a p-basis of $K(X)$.
3. Every extension of a perfect field is separable.

Let $K \subseteq L \subseteq M$. If L is separable over K and M is separable over L, then M is separable over K. If $K \subseteq M$ is separable then so is $K \subseteq L$ but $L \subseteq M$ may not be separable (e.g., $K \subseteq K(x^p) \subseteq K(x)$ for x transcendental over K). The compositum of two separable extensions need not be separable, but it is separable if the two extensions are linearly disjoint.

Using transitivity, an extension of K of the form $K(x_1,...,x_n,x_{n+1})$, where the $x_1,...,x_n$ are algebraically independent over K and x_{n+1} is separably algebraic over $K(x_1,...,x_n)$, is separable. Conversely, every finitely generated separable extension is of this form, as the following theorem says:

Theorem 1.1 (Separating transcendence basis theorem) *If the extension $K \subseteq K(y_1,...,y_n)$ is separable, there exist $m \leq n$ and $i_1 < ... < i_m \leq n$ such that $y_{i_1},...,y_{i_m}$ are algebraically independent over K and $K(y_1,...,y_n)$ is separably algebraic over $K(y_{i_1},...,y_{i_m})$.*

1.2 Note that algebraic extensions never increase the degree of imperfection, and that purely transcendental extensions increase it by the transcendence degree of the extension. Conversely, if K is perfect and B p-free in $K(B)$, then B is algebraically free over K.

Fact 1.3 *Let K be an algebraically closed field and $L = K(x_1, ..., x_n)^s$. Then $K = L^{p^\infty} := \cap_{n \in \mathbb{N}^*} L^{p^n}$.*

Proof: Because K is perfect, $K \subseteq L$ is separable, and we can extract from $x_1, ..., x_n$ a separating transcendence basis, say $x_1, ..., x_m$. As a separable extension of $K(x_1, ..., x_m)$, L has also imperfection degree m. By 1.2, since L^{p^∞} is a perfect subfield of L, the transcendence degree of L over L^{p^∞} is at least m. But L^{p^∞} contains K which is algebraically closed, so $L^{p^\infty} = K$. □

Fact 1.4 *Let $K \subseteq L$ be a separable extension. Then K and L^{p^∞} are linearly disjoint over K^{p^∞}.*

Proof: Let $l_1, ..., l_n \in L^{p^\infty}$ be linearly dependent over K. We have to show they are remain dependent over K^{p^∞}. It suffices to consider the case where every proper subset $\{l_{i_1}, ..., l_{i_{n-1}}\}$ is linearly free over K. Let $k_i \in K$ be such that $\sum k_i l_i = 0$. Each $k_i \neq 0$, and by taking $k_1 = 1$, we get that $k_2, ..., k_n$ are uniquely determined. Hence it is enough to prove they lie in K^{p^r} for every integer r. Since L is a separable extension of K, the fields $K^{p^{-r}}$ and L are linearly disjoint over K, therefore K and L^{p^r} are linearly disjoint over K^{p^r}. This, together with the uniqueness of the k_i's, implies that these k_i's lie in K^{p^r}. □

2 Separably closed fields. Theories and types

Most of the results in this section come from [Er], [Wo 79], [De 88] or [Me 94]. Many of them can also be found in [Me 96]. We give here a slightly different presentation, centered on types.

The theory SC of separably closed fields is axiomatizable in the language of rings: K is separably closed iff each separable polynomial f over K has a root in K. Its completions are
$AC_0 = SC +$ (char=0), and
$SC_{p,\nu} = SC +$ (char=p) + (imperfection degree = ν),
for each prime p and $\nu \in \mathbb{N} \cup \{\infty\}$. We will prove below the completeness of $SC_{p,\nu}$ for finite $\nu > 0$ and $p > 0$, and AC_0 and $SC_{p,0}$ are the theories of algebraically closed fields of given characteristic, and are known to be complete.

From now on, we fix $p > 0$ and ν finite $\neq 0$.

Theorem 2.1 *Each theory $SC_{p,\nu}$ is complete.*

Proof: When studying inclusion of one model in another, we are interested in elementary extensions, hence in our case separable extensions. Because ν is finite, a p-basis of K is still a p-basis of any $L \succeq K$. This justifies adding to the language constants for the elements of a p-basis. Let us prove that in the language $\{0, 1, +, -, .\} \cup \{b_1, ..., b_\nu\}$, the theory $SC_{p,\nu}$ + "$\{b_1, ..., b_\nu\}$ is a p-basis", axiomatized as $\forall x\ (\exists! x_j)_{j \in p^\nu}, x = \sum x_j^p m_j(b_1, ..., b_\nu)$, is model-complete and has a prime model. This will prove completeness [ChKe, 3.1.9].

By 1.2 $b_1, ..., b_\nu$ are algebraically independent over $\mathbb{F}_p$, hence the field $\mathbb{F}_p(b_1, ..., b_\nu)^s$ is uniquely determined and embeds in every model.

Now, by Claim 2.2 below, any model is existentially closed in any model extension, this proves the model-completeness [ChKe, 3.1.7]. □

Claim 2.2 *Let $K \models SC_{p,\nu}$ and let L be a separable extension of K. Then L K-embeds in some elementary extension of K.*

Proof: It is enough to prove it for L finitely generated over K. By 1.1 such an L admits a separating transcendence basis $l_1, ..., l_n$ over K. But any $|K|^+$-saturated elementary extension K^* of K has infinite transcendence degree over K, therefore $K(l_1, ..., l_n)$ K-embeds in K^*, and $K(l_1, ..., l_n)^s$ also since K^* is a model, hence so does L. □

Theorem 2.3 *(1) In the language $\mathcal{L}_{p,\nu} = \{0, 1, +, -, .\} \cup \{b_1, ..., b_\nu\} \cup \{f_i; i \in p^\nu\}$, the theory*

$$T_{p,\nu} = SC_{p,\nu} \cup \{\{b_1, ..., b_\nu\} \text{ is a } p\text{-basis }\} \cup \{x = \sum_{i \in p^\nu} f_i(x)^p m_i(b_1, ..., b_\nu)\}$$

has elimination of quantifiers.
(2) $T_{p,\nu}$ is stable not superstable.

These two results will follow from the description of types of $T_{p,\nu}$ given below.

Let $K \preceq L \models T_{p,\nu}$, with L $|K|^+$-saturated, $x \in L$. $B = \{b_1, \ldots, b_\nu\}$ is a p-basis of L, hence L contains all components x_j, $j \in p^\nu$, of x over B, as well as the components $x_{j \frown k}, k \in p^\nu$, of each x_j, and so on. We index the tree which branches p^ν times at each level by

$$p^\infty := \cup_{n \in \omega} p^{\nu n}$$

where each $p^{\nu n}$ is therefore understood as $(p^\nu)^n$ and one takes a disjoint union. We define now f_j, for $j \in p^\infty$, by setting $f_\emptyset := id$, $f_j := f_{j(n)} \circ \ldots \circ f_{j(1)}$ if $j \in p^{\nu n}$ with $n \geq 1$, and $x_j := f_j(x)$. But $p^{\nu n}$ should also be understood as $\{0, 1, ..., p^n - 1\}^\nu$, when we write

$$x = \sum_{j \in p^{\nu n}} x_j^{p^n} \Pi_{i=1}^\nu b_i^{j(i)}.$$

Lemma 2.4 *$K\langle x\rangle := K(x_i; i \in p^\infty)^s$ is a prime model over $K \cup \{x\}$. It is algebraic (in the model-theoretic sense) over $K \cup \{x\}$.*

Proof: Clearly L contains $K\langle x\rangle$. Now $K \preceq K\langle x\rangle$: we already know that the extension $K \subseteq K\langle x\rangle$ is separable; it remains to prove that B still p-generates $K\langle x\rangle$. This holds, because $K\langle x\rangle$ contains all iterated p-components of x, and hence also all iterated p-components of every $y \in K\langle x\rangle$, by the following lemma. □

Lemma 2.5 *Define* $x_{\leq n} = (x_i; i \in \cup_{m\leq n} p^{\nu m})$. *If* $K \subseteq K(\overline{z}) \subseteq L$ *and* y *is separably algebraic over* $K(\overline{z})$, *then there is a non zero* $d(\overline{z}) \in K[\overline{z}]$ *such that, for all integer* n, $y_{\leq n} \subseteq K[\overline{z}_{\leq n}, d(\overline{z})^{-1}, y]$. *In particular* $y_{\leq n} \subseteq K(\overline{z}_{\leq n}, y)$, *hence each term in the variables* $\overline{x}$ *is equivalent to a rational function in* $\overline{x}_{\leq n}$, *for some integer* n.

Proof: Because $K(\overline{z}) \subseteq K(\overline{z}, y^p) \subseteq K(\overline{z}, y)$ and y is separable over $K(\overline{z})$, $K(\overline{z}, y) = K(\overline{z}, y^p)$. Hence $y \in K[\overline{z}, d(\overline{z})^{-1}, y^p]$ for some $d(\overline{z}) \in K[\overline{z}]$. By iteration, $y \in K[\overline{z}, d(\overline{z})^{-1}, y^{p^n}]$ for each integer n. Now as

$$d(\overline{z})^{-1} = \frac{d(\overline{z})^{p^n-1}}{d(\overline{z})^{p^n}},$$

we get that $y \in (K[\overline{z}_{\leq n}, d(\overline{z})^{-1}, y])^{p^n}(B)$. □

We want to describe the type of x over K, i.e. the isomorphism type of the field $K\langle x\rangle$ over K. We know what a separable closure is, hence we have to describe $K(x_i; i \in p^\infty)$. For this purpose, let us consider the ring

$$K[X_\infty] := K[X_i; i \in p^\infty],$$

where the X_i's are indeterminates. This ring is a countable union of Noetherian rings, hence each ideal is countably generated. We associate to x the following ideal of $K[X_\infty]$

$$I(x, K) := \{f \in K[X_\infty]; f(x_\infty) = 0\}$$

($x_\infty := (x_i; i \in p^\infty)$, we will also sometimes write $f(x)$ for $f(x_\infty)$). In order to describe the range of this map, let us give some definitions.

Definitions and notation:

1. All rings and algebras will be commutative with unit.
2. An ideal I of a K-algebra C is <u>separable</u> if, for all $f_j \in C$, $j \in p^\nu$,

$$\sum_j f_j^p m_j \in I \Rightarrow \text{ each } f_j \in I,$$

where as previously $m_j = m_j(b_1, ..., b_\nu)$. Note that, given a prime ideal I of C, I is separable iff the quotient field of C/I is a separable extension of K. As an intersection of separable ideals is separable, we can speak of the <u>separable closure</u> of some ideal I, which is the smallest separable ideal containing I.

3. For $n \in \omega$,

$$K[X_{\leq n}] := K[X_i; i \in \cup_{m\leq n} p^{\nu m}],$$

and for an ideal I of $K[X_\infty]$

$$I_{\leq n} := I \cap K[X_{\leq n}].$$

Note that I is separable (or prime) iff each $I_{\leq n}$ is.
4. Let I^0 be the ideal of $K[X_\infty]$ generated by the polynomials $X_i - \sum_{j \in p^\nu} X^p_{i\frown j} m_j$, $i \in p^\infty$.

The following lemma will be used further on in section 5.

Lemma 2.6 *In an algebra over a separably closed field, any prime separable ideal is absolutely prime.*

Proof: If I is a prime separable ideal of the K-algebra C, the quotient field $Q(C/I)$ of C/I is a separable extension of K. Since the extension $K \subseteq K^a$ is purely inseparable, K^a and $Q(C/I)$ are linearly disjoint over K. □

Proposition 2.7 *The map $x \to I(x, K)$ defines a bijection between 1-types over K and prime separable ideals I of $K[X_\infty]$ containing I^0.*

Proof: This map clearly defines an injection. Consider now such an ideal I. Let M be the quotient field of $K[X_\infty]/I$. Then $K \preceq M^s$. For $x := X/I$, $I = I(x, K)$. □

The k-types are described as well, as L^k embeds in L via the following map:

$$(x_{(0)}, \ldots, x_{(k-1)}) \to \sum_{i \in p^{\nu n}} x^{p^n}_{(i)}.m_i$$

for n such that $p^{\nu n} \geq k$ and $x_{(k)} = \ldots = x_{(p^{\nu n}-1)} = 0$ (here $p^{\nu n}$ is regarded both as an integer and as the set $\{0, 1, \ldots, p^{\nu n} - 1\}$). If we define in the same way $K[X_{1\infty}, \ldots, X_{k\infty}] := K[X_{1i}, \ldots, X_{ki}; i \in p^\infty]$ and for $x \in L^k$, $I(x, K) := \{f \in K[X_{1\infty}, \ldots, X_{k\infty}]; f(x_\infty) = 0\}$, and if $I^0(X_i).K[X_{1\infty}, \ldots, X_{k\infty}]$ is simply denoted as $I^0(X_i)$, then

Proposition 2.8 *The map $x \to I(x, K)$ defines a bijection between k-types over K and prime separable ideals I of $K[X_{1\infty}, \ldots, X_{k\infty}]$ containing $\sum_{i=1}^k I^0(X_i)$.*

Definition: A prime ideal of $K[X_{1\infty}, \ldots, X_{k\infty}]$, separable and containing $\sum_{i=1}^k I^0(X_i)$, will be called a type ideal.

We now describe types over sets and prove quantifier elimination. The reader who is willing to admit this result can proceed directly to 2.11. A direct proof can also be found in [Me 96]. It is enough to consider definably closed sets. These are very close to being models, as the following lemma says.

Lemma 2.9 *Let B be the p-basis and $A \subseteq K$.*
(1) $\langle A\rangle := \mathbb{F}_p(B, f_j(A); j \in p^\infty)^s$ is a prime model over A.
(2) The definable closure of A is $(A)_{df} := \mathbb{F}_p(B, f_j(A); j \in p^\infty)$. Equivalently, A is definably closed iff it is a field, containing the p-basis B and closed under the f_i's, $i \in p^\nu$.
(3) $(A)_{df}$ is quantifier free definable over A in $\mathcal{L}_{p,\nu} \cup \{^{-1}\}$.

Proof: Clear once noted that no point of $\langle A\rangle \setminus (A)_{df}$ is definable over A since it can be moved by some A-automorphism. □

Thus a definably closed subset A of K is a subfield containing B as a p-basis and we can define as previously the rings $A[X_{1\infty}, ..., X_{k\infty}]$ and $A[X_{1\leq n}, ..., X_{k\leq n}]$, separable ideals in them, $I_{\leq n}$ for an ideal I of $A[X_{1\infty}, \dots, X_{k\infty}]$, and for $x \in L^k$,

$$I(x, A) := \{f \in A[X_{1\infty}, ..., X_{k\infty}]; f(x_\infty) = 0\}.$$

We will still denote by $I^0(X_i)$ the ideal $I^0(X_i) \cap A[X_{1\infty}, ..., X_{k\infty}]$.

Then we can state:

Proposition 2.10 *For every integer k, k-types over a definably closed set of parameters A are in bijection with prime separable ideals I of $A[X_{1\infty}, \dots, X_{k\infty}]$ containing $\sum_{i=1}^k I^0(X_i)$. All extensions to $\langle A\rangle$ are non forking and conjugate over A.*

Proof: A type P over A has only non forking extensions to $\langle A\rangle$ because $\langle A\rangle$ is algebraic over A. If $I := I(x, A)$ for some realization x of P, I is clearly prime, contains $\sum_{i=1}^k I^0(X_i)$ and is separable since $I(x, K)$ is. Conversely, for such an I, by classical results over Noetherian polynomial rings, the minimal prime ideals of $A^s[X_{1\leq n}, ..., X_{k\leq n}]$ containing $I_{\leq n} \otimes A^s$ are conjugate and intersect $A[X_{1\leq n}, ..., X_{k\leq n}]$ in $I_{\leq n}$. Now, by considering the dimension, we see that any prime ideal of $A^s[X_{1\leq n}, ..., X_{k\leq n}]$ intersecting $A[X_{1\leq n}, ..., X_{k\leq n}]$ in $I_{\leq n}$ is minimal over $I_{\leq n} \otimes A^s$. Therefore, the various ideals of $A^s[X_{1\infty}, \dots, X_{k\infty}]$ intersecting $A[X_{1\infty}, \dots, X_{k\infty}]$ in I are conjugate. Let Q be one such ideal. Then the quotient field of $A^s[X_{1\infty}, \dots, X_{k\infty}]/Q$ is a composite of A^s and of a subfield L which is A-isomorphic to the quotient field of $A[X_{1\infty}, \dots, X_{k\infty}]/I$. Since L is a separable extension of A, and A^s is separably algebraic over A, their composite is also separable over A, and therefore over A^s. This shows that Q is separable, hence is the ideal of a type over A^s. □

Proof of Theorem 2.3 :
(1) In the language $\mathcal{L}_{p,\nu} \cup \{^{-1}\}$, the definable closure of any set of parameters is quantifier free definable and quantifier free types over definably closed sets of parameters are complete. This implies quantifier elimination in $\mathcal{L}_{p,\nu} \cup \{^{-1}\}$. By 2.5 (or by an easy induction on the complexity of terms) any term of $\mathcal{L}_{p,\nu} \cup \{^{-1}\}$

is of the form uv^{-1}, where u and v are terms of $\mathcal{L}_{p,\nu}$. Hence any atomic formula of $\mathcal{L}_{p,\nu} \cup \{^{-1}\}$ is equivalent to some atomic formula of $\mathcal{L}_{p,\nu}$.
(2) $I(x, K)$ is countably generated, hence $|S_1(K)| \leq |K|^\omega$, which proves the stability. Now, in the sequence

$$K \supset K^p \supset \ldots \supset K^{p^n} \supset K^{p^{n+1}} \supset \ldots$$

of additive subgroups of K, each $K^{p^{n+1}}$ has infinite index in K^{p^n} as K^{p^n} is also a $K^{p^{n+1}}$-vector space and $K^{p^{n+1}}$ is infinite. This contradicts superstability. □

Proposition 2.11 *1. Each type $t \in S_1(K)$ has a countable field of definition D (: $\Leftrightarrow$ for any saturated $K^* \succeq K$ and any automorphism σ of K^*, σ preserves the non forking extension of t over K^* iff $\sigma|D = id_D$).*
2. For $K \preceq L \preceq F$ and $x \in F$, $t(x, L)$ does not fork over K iff $I(x, L) = L \otimes I(x, K)$ iff $t(x, L)$ has a field of definition contained in K iff L and $K\langle x\rangle$ are linearly disjoint over K.

For the proof see [De 88].

Some remarks about ranks and generics:
1. $I^0 = I(t, K)$ for t generic over K. (We mean here generic in the sense of the theory of stable groups; this notion has only been defined in this volume for the case of ω-stable groups but the reader can take the previous statement as a definition of "generic over K".)
2. By non superstability, the generic can not be U-ranked. We can see this directly: if x is generic over K and

$$K_n := K\langle x_0, x_{10}, \ldots, x_{1\ldots 10}\rangle,$$

then $t(x, K_{n+1})$ forks over K_n. 3. In Section 5, we give a precise analysis of minimal types, i.e. of types with U-rank 1. In [De 88, 49], an algebraic interpretation of finite U-rank is given.
4. Any non algebraic formula contains a point having some generic p^n-component: see the remark following 3.4. Hence there is no non trivial continuous rank.

Theorem 2.12 *$T_{p,\nu}$ has elimination of imaginaries.*

Proof in the next section, see [Zie] for the definition of elimination of imaginaries.

We are now entitled to use all the machinery of stability in the context of separably closed fields. This enables us to characterize the groups interpretable in $T_{p,\nu}$ (in analogy to Weil's theorem, see [Pil, 4.12] in this volume) and to describe the interpretable fields [Me 94]. The proofs of these results use techniques

In the present paper $\subset$ and $\supset$ always denote strict inclusion.

from geometric stability theory. From this work we will quote only result 2.13 below.

For $a_1, \ldots, a_m \in K \models T_{p,\nu}$, $K^{p^n}[a_1, \ldots, a_m]$ is clearly definable in K. Conversely :

Theorem 2.13 *An infinite field interpretable in $K \models T_{p,\nu}$ is definably isomorphic to a subfield $K^{p^n}[a_1, \ldots, a_m]$ of K.*

Proposition 2.14 *For $K \models T_{p,\nu}$, the field K^{p^∞} is algebraically closed, it is the largest algebraically closed subfield of K. It is $\bigwedge$-definable in K and, as such, is a pure field, which means that for every $F \subseteq K^k$ definable in K with parameters from K, $F \cap (K^{p^\infty})^k$ is definable in the field K^{p^∞} with parameters from K^{p^∞}.*

Proof: The field K^{p^∞} is separably closed since it is the intersection of separably closed fields (each K^{p^n} being isomorphic to K). It is also perfect, hence algebraically closed. It is clearly the largest algebraically closed subfield of K. By quantifier elimination any formula $\phi(x_1, \ldots, x_k, \overline{c})$ of $\mathcal{L}_{p,\nu}$ is a Boolean combination of equations

$$f(x_{1\leq n}, \ldots, x_{k\leq n}, \overline{c}_{\leq n}) = 0$$

for some integer n and some $f \in \mathbb{F}_p[X_{1\leq n}, \ldots, X_{k\leq n}, \overline{C}_{\leq n}]$. For $x \in K^{p^\infty}$, the p^n-components of x are all zero, except the one corresponding to the p^n-monomial 1, which is equal to $x^{p^{-n}}$, hence quantifier free definable in the ring language. Now, as the trace over a subfield of a Zariski closed set is Zariski closed in the small field, we get that, for $x_1, \ldots, x_k \in K^{p^\infty}$, $\phi(x_1, \ldots, x_k, \overline{c})$ is equivalent to a formula of the ring language in $x_1, \ldots, x_k$ with parameters from K^{p^∞} (one can also use directly the fact that in a stable theory if A is infinitely definable in a model M, then any definable subset of A is definable with parameters from A). □

Proposition 2.15 *Let $F \subseteq K^k$ be definable with parameters from K, and $h : F \rightarrow K$ be a map definable with parameters from K. Then there exist $C_1, \ldots, C_m \subseteq (K^{p^\infty})^k$ definable in the field K^{p^∞} with parameters from K^{p^∞} and such that*

- $F \cap (K^{p^\infty})^k = C_1 \cup \ldots \cup C_m$ and

- each $h \upharpoonright C_i$, for $i = 1, \ldots, m$, is a composition of rational functions and of the inverse of the Frobenius (these functions $h \upharpoonright C_i$'s may have parameters from K).

Proof: By quantifier elimination and compactness there are an integer n and definable $D_1, \ldots, D_m \subseteq K^k$ such that each $h \upharpoonright D_i$ is a rational function in $x_{1\leq n}, \ldots, x_{k\leq n}$ (the proof is along the same lines as the similar statement for algebraically closed fields, see [Pil, 1.5] this volume). By 2.14 each $D_i \cap (K^{p^\infty})^k$ is definable in K^{p^∞} and, arguing as in the proof of 2.14, for $x \in K^{p^\infty}$, all terms of the sequence $x_{\leq n}$ are zero except $x^{p^{-n}}$. □

Definition: An infinite $\bigwedge$-definable subset A is minimal if the trace on A of any definable subset is finite or cofinite in A. It follows trivially from 2.14 that the $\bigwedge$-definable field K^{p^∞} is minimal. Conversely :

Proposition 2.16 *An infinite field k which is $\bigwedge$-interpretable in $K \models T_{p,\nu}$ and minimal is definably isomorphic to K^{p^∞}.*

Proof: By [Hr 90], the stability of $T_{p,\nu}$ implies that there exist a field k^* interpretable in K and definable subfields $(k_n)_{n\in\omega}$ of k^* such that $k = \bigcap_{n\in\omega} k_n$. By 2.13, k^* is definably isomorphic to a subfield $l := K^{p^n}[a_1, ..., a_m]$. Via this isomorphism, each k_n becomes a subfield l_n of l containing K^{p^∞}. Hence $\bigcap l_n \supseteq K^{p^\infty}$. As a minimal field is algebraically closed, and K^{p^∞} is the largest algebraically closed field contained in K, $\bigcap l_n = K^{p^\infty}$. □

Remark: Decidability and stability of separably closed fields with infinite imperfection degree are proved along the same lines. It is also possible to describe the types and to give a natural language eliminating quantifiers in this setting. But we do not know any language eliminating imaginaries. Thus we are unable to characterize the interpretable groups and fields.

3 λ-closed subsets of affine space

Let us fix $K \preceq L \models T_{p,\nu}$. We will consider some particular subsets of L^k which are $\bigwedge$-definable with parameters from K.
Recall that if $x \in L^k$ and $f \in K[X_{1\infty}, \ldots, X_{k\infty}]$, we allow ourselves to write $f(x)$ for $f(x_\infty)$ (or more accurately for $f(x_{1\infty}, \ldots, x_{k\infty})$).

Definition: Given a set of polynomials S of $K[X_{1\infty}, \ldots, X_{k\infty}]$, we define

$$V(S) = \{x \in L^k;\ \text{each polynomial of } S \text{ vanishes on } x\}$$

("for all $f \in S, f(x) = 0$" will also be denoted "$S(x) = 0$"). Such a $V(S)$ is called λ-closed (with parameters in K) in L^k. It is defined by a countable conjunction of first-order formulas (each ideal of $K[X_{1\infty}, \ldots, X_{k\infty}]$ is countably generated).

The following properties are clear.

Proposition 3.1 *For ideals I, J and I_α of $K[X_{1\infty}, \ldots, X_{k\infty}]$, we have*

$$V(I \cdot J) = V(I) \cup V(J) = V(I \cap J)$$
$$V(\sum_\alpha I_\alpha) = \bigcap_\alpha V(I_\alpha)$$
$$V(K[X_{1\infty}, \ldots, X_{k\infty}]) = \emptyset$$
$$V(0) = L^k.$$

(Note that the equality $V(I) \cup V(J) = V(I \cap J)$ does not generalise to infinite intersections and unions.) As a corollary:

Proposition 3.2 *The λ-closed sets are the closed sets of some topology over L^k.*

Remark: In this topology, as in the classical Zariski topology on L^k, the points of K^k are the unique separated points of L^k, hence it is T_0 iff $K = L$. On the other hand, it is not Noetherian, as the following sequence (of λ-closed sets of L) shows:
$V(X_0) \supset V(X_0, X_{10}) \supset V(X_0, X_{10}, X_{110}) \supset ... \supset V(X_0, X_{10}, \ldots, X_{11...10}) \supset ...$
(X is here a single variable, and the indices describe its iterated p-components).

Definition: Given $A \subseteq L^k$, we define its <u>canonical ideal</u> $I(A, K)$, or $I(A)$ when there is no ambiguity,

$$I(A) := \{f \in K[X_{1\infty}, \ldots, X_{k\infty}];\ \forall a \in A,\ f(a) = 0\}$$

("$\forall a \in A,\ f(a) = 0$" will also be denoted "$f(A) = 0$"). In particular we write now $I(x)$ for the ideal previously denoted as $I(x, K)$.

Remarks:
1. $I(\cup A_\alpha) = \cap I(A_\alpha)$.
2. $V(I(A))$ is the closure of A in the topology defined above, or λ-<u>closure</u>.

From now on in this section, L is ω_1-saturated.

Proposition 3.3 ("Nullstellensatz") 1. *The map $A \to I(A)$ defines a bijection between λ-closed subsets of the affine space L^k with parameters in K, and ideals of $K[X_{1\infty}, \ldots, X_{k\infty}]$ which are separable and contain $\sum_{i=1}^k I_i^0$. The inverse map is $I \to V(I)$.*
2. *An ideal of $K[X_{1\infty}, \ldots, X_{k\infty}]$ is of the form $I(A)$ iff it is the intersection of all type ideals containing it.*

Consequently, given an ideal I of $K[X_{1\infty}, \ldots, X_{k\infty}]$, $I(V(I))$ is the separable closure of the ideal $I + \sum I_i^0$.

The proof of 3.3 will use the following facts.

Notation: For an ideal I of $K[X_{1\infty}, \ldots, X_{k\infty}]$,

$$I_{\leq n} := I \cap K[X_{1\leq n}, ..., X_{k\leq n}].$$

Note that the condition "$I_{\leq n}(x) = 0$" is first-order in x.

Fact 3.4 *Let q be an ideal of $K[X_{1\leq n}, ..., X_{k\leq n}]$, prime, separable and containing $\sum_{i=1}^k I_{i\leq n}^0$. Then there exists a k-type P satisfying $I(P)_{\leq n} = q$.*

Proof: Since q is prime separable, by Claim 2.2, the fraction field of $K[X_{1\leq n},...,X_{k\leq n}]/q$ K-embeds in some elementary extension of K. Let x_{ij}, for $i=1,...,k$ and $j\in\cup_{m\leq n}p^{\nu m}$, be the images of $X_{ij}+q$ under this embedding. As $q\supseteq\sum_{i=1}^{k}I_i^0$, the x_{ij}'s, $j\in\cup_{m\leq n}p^{\nu m}$, are the iterated p-components of $x_i(=x_{i\emptyset})$. Take now $P:=t(x_1,...,x_k;K)$. □

Remark: One can prove the following more precise fact: there exists a unique type ideal $I(P)$ intersecting $K[X_{1\leq n},...,X_{k\leq n}]$ in q and minimal for inclusion. If r is the Krull dimension of q, the type P has r p^n-components which are independent realizations of the generic. The proof is in the same spirit as the proof of 5.3.(3).

Fact 3.5 *A separable ideal of a K-algebra is radical.*

Proof: If a separable ideal contains some power x^n, it contains x^{p^m} for all $p^m\geq n$ and hence also x by separability. □

Fact 3.6 *Let I,Q,J be ideals. Suppose that $I=Q\cap J$ is separable, $J\not\subseteq Q$ and Q is prime. Then Q is separable.*

Proof: Suppose that $f=\sum_j f_j^p m_j\in Q$; choose $g\in J\setminus Q$. Then $g^pf\in I$, and each $gf_j\in I$ by separability of I. Since Q is prime, this implies that $f_j\in Q$. □

Proof of 3.3 : 1. Using the relations

$$\begin{aligned} I(V(I(A))) &= I(A)\\ V(I(V(I))) &= V(I), \end{aligned}$$

we see that A is λ-closed iff $A=V(I(A))$, and that $A\to I(A)$ and $I\to V(I)$ define reciprocal bijections between λ-closed sets and ideals of the form $I(A)$ for $A\subseteq L^k$. An ideal of the form $I(A)$ is clearly separable and contains $\sum I_i^0$. Conversely, let I be an ideal of $K[X_{1\infty},\ldots,X_{k\infty}]$, separable and containing $\sum_i I_i^0$, and $g\in K[X_{1\leq n},...,X_{k\leq n}]\setminus I$. Then for all integer m

$$g\in K[X_{1\leq n+m},\ldots,X_{k\leq n+m}]\setminus I_{\leq n+m}.$$

By classical results on commutative rings, since $I_{\leq n+m}$ is radical there exists a prime ideal q of $K[X_{1\leq n+m},...,X_{k\leq n+m}]$ containing $I_{\leq n+m}$ but not g. We may choose this q to be minimal prime over $I_{\leq n+m}$. By 3.6 q is separable and by 3.4 there is some type P satisfying $I(P)_{\leq n+m}=q$, hence $g\notin I(P)$. On a point realizing P, $I_{\leq n+m}$ vanishes and not g, that is, if $K\langle P\rangle$ denotes the prime model over K and some realization of P,

$$K\langle P\rangle\models\exists x\ I_{\leq n+m}(x)=0\wedge g(x)\neq 0.$$

K and L must satisfy the same formula, and then, by ω_1-saturation, L satisfies

$$\exists x\ I(x)=0\wedge g(x)\neq 0.$$

This proves $g \notin I(V(I))$, and finally $I(V(I)) = I$.

2. An intersection of type ideals is clearly separable and contains $\sum_i I_i^0$. Conversely a λ-closed set A is $\wedge$-definable, hence

$$A = \cup\{ \text{ realizations of } P \text{ in } L^k;\ P \text{ complete } k\text{-type} \vdash x \in A\},$$

Therefore

$$I(A) = \cap\{I(P);\ P \text{ complete } k\text{-type} \vdash x \in A\}. \qquad \square$$

Proposition 3.7 *Ideals of λ-closed sets are stable under sum, i.e. $I(A \cap B) = I(A) + I(B)$ for A and B λ-closed.*

This proposition is slightly surprising because there is no analogous result in algebraic geometry, the sum of two radical ideals not being in general a radical ideal. The difference here comes from the fact that the polynomial rings we are working with are not Noetherian. The ideals we are considering, which have to contain $\sum_{i=1}^k I_i^0$, are resolutely of non finitary type. In particular, $(I + J)_{\leq n}$ will in general strictly contain $I_{\leq n} + J_{\leq n}$, which need not be a separable ideal of $K[X_{1 \leq n}, ..., X_{k \leq n}]$.

By 3.3, it suffices to show that $I(A) + I(B)$ is separable. Thus, the result follows immediatly from:

Lemma 3.8 *Let I and J be ideals of $K[X_{1\infty}, \ldots, X_{k\infty}]$ separable and containing $\sum_{i=1}^k I_i^0$. Then $I + J$ is separable.*

Proof: Let $\sum x_j^p m_j = a + b$, with $x_j, a, b \in K[X_{1\leq n}, ..., X_{k\leq n}], a \in I, b \in J$, and write

$$a \equiv \sum a_j^p m_j \ (\text{mod} \sum I_i^0),$$

$$b \equiv \sum b_j^p m_j \ (\text{mod} \sum I_i^0),$$

with

$$a_j, b_j \in K[X_{1\leq n+1}, ..., X_{k\leq n+1}].$$

Since I and J are separable and contain $\sum I_i^0$, $a_j \in I$ and $b_j \in J$. Now

$$\sum (x_j - a_j - b_j)^p m_j \in \sum I_i^0,$$

and $\sum I_i^0$ is separable, hence

$$x_j - a_j - b_j \in \sum I_i^0.$$

Therefore $x_j \in I + J$. $\square$

Proposition 3.9 *$T_{p,\nu}$ has elimination of imaginaries.*

Proof: We will prove that any definable subset D of some K^k has a field of definition, which means that there is a field $K_0 \subseteq K$ such that, for $K^* \succeq K$ and any automorphism σ of K^*, the canonical extension of D over K^* is invariant under σ iff $\sigma|K_0 = \mathrm{id}_{K_0}$. By quantifier elimination, D is defined by a formula

$$\vee_i(\wedge_j P_{i,j}(x) = 0 \wedge Q_i(x) \neq 0)$$

with $P_{i,j}, Q_i \in K[X_{1\infty}, \ldots, X_{k\infty}]$. Now, σ preserves D iff it preserves the formula in $x^\frown y$, where $y = (y_i)_i$,

$$\vee_i(\wedge_j P_{i,j}(x) = 0 \wedge Q_i(x).y_i = 1).$$

Hence it is enough to prove the result for D λ-closed, which follows by 3.3 from the existence of the field of definition of an ideal. □

Definitions:
1. In a topological space, a closed set is <u>irreducible</u> if it is not the proper union of two closed subsets.
2. A maximal irreducible closed subset of some closed set A is called an <u>irreducible component</u> of A.
3. A point of some irreducible closed subset A is called <u>generic</u> in A if its closure is A.

Proposition 3.10 *In an arbitrary topological space,*
1. a closed set is the union of its closed irreducible subsets;
2. if a closed set A is the union of finitely many maximal irreducible closed sets A_i, then the A_i's are all the irreducible components of A.

Proof: For 1. note that the closure of a singleton is irreducible, and that a closed set is the union of closures of its points. For 2. let $A = F_1 \cup \ldots \cup F_n \supseteq F$, with $F, F_1, \ldots, F_n$ irreducible closed sets. For $i = 1, \ldots, n$, $F = (F_i \cap F) \cup (\cup_{j \neq i} F_j \cap F)$, and by irreducibility of F, either $F \subseteq F_i$ or $F \subseteq \cup_{j \neq i} F_j$. An induction shows that some F_i contains F. □

Proposition 3.11 *A λ-closed set A is irreducible iff $I(A)$ is prime.*

Proof: Consider three λ-closed subsets $A = B \cup C$. Then $I(A) = I(B) \cap I(C) \supseteq I(B) \cdot I(C)$. If $A \supset B, C$, then $I(A) \subset I(B), I(C)$ and there exist $f \in I(B) \setminus I(A)$ and $g \in I(C) \setminus I(A)$, so $fg \in I(A)$. Hence $I(A)$ is not prime if A is reducible. Conversely if $I(A)$ is not prime, there are f and $g \in K[X_{1\infty}, \ldots, X_{k\infty}] \setminus I(A)$ such that $fg \in I(A)$. Then

$$B := \{a \in A; f(a) = 0\}, \text{ and}$$

$$C := \{a \in A; g(a) = 0\}$$

are proper λ-closed subsets of A, and $A = B \cup C$. □

Remarks:

1. By 3.11, a point x of an irreducible λ-closed set A is generic in A iff $I(x) = I(A)$. We can also give the following interpretation. We just proved that A is irreducible iff $I(A)$ is prime, therefore there is some k-type P satisfying $I(P) = I(A)$. Thus the generic points of A are exactly the realizations of P. Note that P is strictly included in A unless A is a singleton, in which case P is realized.

2. There is a priori no reason why 3.10.2 should remain true when A has infinitely many components. The dual problem over ideals is as follows. We know that

$$I(A) = \cap\{I(P); P \text{ } k\text{-type completing } A\}.$$

We can partly reduce the intersection and write

$$I(A) = \cap\{Q; Q \text{ prime separable ideal } \supseteq I(A) \text{ and minimal for this property}\},$$

as a decreasing intersection of prime separable ideals is again prime separable. Is such an intersection reduced ? (An intersection $\cap_{\alpha<\alpha_0} Q_\alpha$ is reduced if for all $\alpha_1 < \alpha_0$, $\cap_{\alpha<\alpha_0, \alpha\neq\alpha_1} Q_\alpha$ strictly contains $\cap_{\alpha<\alpha_0} Q_\alpha$.) A priori not. Hence we do not know how well in general these irreducible components behave.

Definition: A set of the form $V(I)$ for a finitely generated I is called λ-closed of finite type (in [Me 94] such sets are called basic λ-closed.)

Remark:
Any λ-closed set is an intersection of λ-closed sets of finite type:

$$x \in V(I) \text{ iff } \wedge_n \, x \in V(I_{\leq n}).$$

And a λ-closed set is definable iff it has finite type. Indeed, by compactness, an infinite conjunction of first-order formulas which expresses a first-order condition is equivalent to a finite sub-conjunction. In other words, quantifier elimination can be restated as follows: a definable set is a finite Boolean combination of λ-closed sets of finite type.

We wish now to make the connection between $V(I)$ and the $I_{\leq n}$'s more precise.

Notation:

1. To x in L^k, we associate (cf. 2.5)

$$x_{\leq n} := (x_i; i \in \cup_{m\leq n} p^{\nu m}) \in K^{k(1+p^\nu+\ldots+p^{\nu n})}.$$

So $x, x_\emptyset$ and $x_{\leq 0}$ are canonically identifiable, and the $x_{\leq n}$'s form a projective system with limit x_∞. For $D \subseteq K^k$, define

$$D_{\leq n} := \{x_{\leq n}; x \in D\} \subseteq K^{k(1+p^\nu+\ldots+p^{\nu n})}.$$

Quantifier elimination implies that for any definable D there is some integer n such that $D_{\leq n}$ is quantifier free definable in the ring language.

2. For $D \subseteq L^{k(1+p^\nu+...+p^{\nu n})}$ and an ideal I of $K[X_{1\leq n},...,X_{k\leq n}]$,

$$I^{(n)}(D) := \{f \in K[X_{1\leq n},...,X_{k\leq n}]; f(D) = 0\},$$

$$V^{(n)}(I) := \{x \in L^{k(1+p^\nu+...+p^{\nu n})}; I(x) = 0\}.$$

If we identify x and x_∞, a set of the form $V(I)$ can be understood as living in the projective limit of the $L^{k(1+p^\nu+...+p^{\nu n})}$, $n \in \omega$, and then $V(I)_{\leq n}$ appears as the projection of a λ-closed set. It is not in general Zariski closed in $L^{k(1+p^\nu+...+p^{\nu n})}$, but we have the following

Lemma 3.12 *Suppose $I = I(V(I))$. Then*

$$I_{\leq n} = I^{(n)}(V^{(n)}(I_{\leq n})) = I^{(n)}(V(I)_{\leq n}).$$

Hence the Zariski closure of $V(I)_{\leq n}$ in $L^{k(1+p^\nu+...+p^{\nu n})}$ is $V^{(n)}(I_{\leq n})$.

Proof: $V^{(n)}(I_{\leq n}) \supseteq V(I)_{\leq n}$, hence $I^{(n)}(V^{(n)}(I_{\leq n})) \subseteq I^{(n)}(V(I)_{\leq n})$. And clearly $I_{\leq n} \subseteq I^{(n)}(V^{(n)}(I_{\leq n}))$. Now, if $f \in K[X_{1\leq n},...,X_{k\leq n}] \setminus I_{\leq n}$, there is some $x \in L^k$ satisfying $I(x) = 0 \wedge f(x) \neq 0$, hence $I^{(n)}(V(I)_{\leq n}) \subseteq I_{\leq n}$. □

4 λ-closed subsets of a fixed type

In this section we relativize the notion of closed set introduced previously to the set of realizations of a single type: we consider only tuples from L^k whose coordinates all realize a fixed one-type over a small set of parameters.

The hypotheses are the following: $K_0 \preceq K \preceq L \models T_{p,\nu}$. The fixed one-type is defined over K_0, the closed sets are defined over K.

We suppose that L is $|K_0|^+$-saturated. Recall from the previous section that a λ-closed set is defined by a countable conjunction of equations.

Notation:

. P is a complete 1-type over K_0,

. $K_0\langle P\rangle$ is the prime model over K_0 and some realization of P,

. $P_i := P(x_i)$,

. P^k is the conjunction $\wedge_{i=1}^k P(x_i)$, which, even on K_0, is incomplete unless $k = 1$ or P is the type of some element in K_0,

. $P_1 \otimes ... \otimes P_k$ is the complete k-type over K_0 of an independent k-tuple of realizations of P, i.e. the type over K_0 corresponding to the ideal $\Sigma I(P_i)$ (following the notation introduced before Proposition 2.8, $I(P_i)$ denotes an ideal of $K[X_{i\infty}]$ as well as of $K[X_{1\infty},\ldots,X_{k\infty}]$).

. For $R \in S_1(K_0)$, $h(R;K)$ is the non forking extension of R to K i.e. the type over K corresponding to the ideal $I(R).K[X_{1\infty},\ldots,X_{k\infty}]$. This ideal is isomorphic to $I(R) \otimes_{K_0} K$ and will sometimes also be denoted as $I(R)$. In such a way, I^0 may denote an ideal of $K'[X_\infty]$ for any $K' \succeq K_0$.

. $P(L)$ is the set of realizations of P in L.

Definition: 1. $A \subseteq P^k$ is λ-closed in P^k, with parameters from K, if it is the trace on P^k of some λ-closed subset, with parameters from K, of the affine space, i.e. $A = V(I) \cap P^k$.

2. If $A = V(I) \cap P^k$, A is λ-closed of finite type in P^k if it is the trace of some λ-closed subset of finite type of the affine space.

Remarks:

1. A possible interpretation of λ-closed subsets of P^k is as follows. Since P is a complete type, for $x = (x_1, \ldots, x_k) \in P^k$, all fields $K_0\langle x_i\rangle$ are K_0-isomorphic to $K_0\langle P\rangle$. The question arises how these k copies of $K_0\langle P\rangle$ relate to each other, and relate to K. If K and all $K_0\langle x_i\rangle$'s are linearly disjoint over K_0 (which is possible, since the extensions $K_0 \subseteq K$ and $K_0 \subseteq K_0\langle x_i\rangle$ are regular), then $K[x_{1\infty}, \ldots, x_{k\infty}] \simeq_K K \otimes_{K_0} K_0[x_{1\infty}] \otimes_{K_0} \ldots \otimes_{K_0} K_0[x_{k\infty}]$ and $I(x, K) = \sum I(P_i) \otimes_{K_0} K = I(h(P_1 \otimes \ldots \otimes P_k; K))$. At the other extreme, the fields $K_0\langle x_i\rangle$ may coincide, for example if $x_1 = x_2 = \ldots = x_k$.
For $A = V(I) \cap P^k$, the condition "$x \in A$" is equivalent to the following

each $K_0\langle x_i\rangle \simeq_{K_0} K_0\langle P\rangle$ and
$K[x_{1\infty}, \ldots, x_{k\infty}]$ is a quotient of $K[X_{1\infty}, \ldots, X_{k\infty}]/I$.

2. For $A = V(I) \cap P^k$, A is the set of points of $V(I + \sum I(P_i))$ having their coordinates generic in the λ-closed set defined (over K_0) by the ideal $I(P)$. Hence A is not λ-closed in the affine space, except when P is realized in K_0.

3. Any λ-closed subset of P^k is $\bigwedge$-definable in L^k over some set of cardinality $|K_0|^+$.

4. A λ-closed subset $A = V(I) \cap P^k$ of P^k is of finite type iff $A = D \cap P^k$ for some D definable in the affine space iff $A = V(I_{\leq n}) \cap P^k$ for some integer n (by compactness applied to : $x \in P^k \vdash x \in D \leftrightarrow \bigwedge_n x \in V(I_{\leq n})$).

Proposition 4.1 *The map $A \to I(A)$ defines a bijection between the non empty λ-closed subsets of P^k and the ideals I of $K[X_{1\infty}, \ldots, X_{k\infty}]$ such that $I = \cap\{Q; Q \in \mathcal{Q}\}$, with a non empty*

$$\mathcal{Q} := \{Q \textit{ an ideal of } K[X_{1\infty}, \ldots, X_{k\infty}]; Q \textit{ is prime, separable, } Q \supseteq I$$

$$\textit{and } Q \cap K_0[X_{i\infty}] = I(P_i) \textit{ for } i = 1, \ldots, k\}$$

(in particular such an ideal I is separable and satisfies $I \cap K_0[X_{i\infty}] = I(P_i)$, for $i = 1, \ldots, k$). The inverse map is $I \to V(I) \cap P^k$.

Proof: If A is a λ-closed subset of P^k, it is $\wedge$-definable, hence

$$I(A) = \cap\{I(R); R \text{ a complete } k\text{-type over } K, R \vdash A\}.$$

Now each $I(R)$ is prime, separable, extends $I(A)$ and, as A is contained in P^k and is not empty, intersects each $K_0[X_{i\infty}]$ in $I(P_i)$. Conversely, let I be some ideal of $K[X_{1\infty}, \ldots, X_{k\infty}]$ satisfying $I = \cap\{Q; Q \in \mathcal{Q}\}$, for $\mathcal{Q} := \{Q; Q$ prime, separable, $Q \supseteq I$ and $Q \cap K_0[X_{i\infty}] = I(P_i)$ for $i = 1, \ldots, k\}$ non empty, and $g \in K[X_{1\leq n}, \ldots, X_{k\leq n}] \setminus I$. Then $g \notin Q_{n+m}$ for some $Q = I(R) \in \mathcal{Q}$ and every integer m, hence

$$K\langle R\rangle \models \exists x\ [Q_{\leq n+m}(x) = 0 \wedge g(x) \neq 0].$$

The same argument as in 3.3, using this time the $|K_0|^+$-saturation of L, shows

$$L \models \exists x\ [I(x) = 0 \wedge \wedge_{i=1}^{k} I(x_i; K_0) = I(P_i) \wedge g(x) \neq 0],$$

in other words

$$L \models \exists x\ [x \in (P^k \cap V(I)) \wedge g(x) \neq 0].$$

It follows that $g \notin I(P^k \cap V(I))$. □

A reducible (respectively irreducible) λ-closed subset of the affine space may have an irreducible (respectively a reducible) trace on P^k, but an irreducible λ-closed subset of P^k is the trace of some irreducible λ-closed subset of the affine space, as we have:

Proposition 4.2 *A λ-closed subset A of P^k is irreducible in P^k iff $V(I(A))$ is irreducible in the affine space iff $I(A)$ is prime.*

Proof: Same proof as for 3.11. □

So, as in the case of affine space, a λ-closed subset A of P^k is irreducible iff $I(A)$ is prime iff there is an $x \in L^k$ such that $I(x) = I(A)$. All such x have the same type, which completes P^k and is the type corresponding to $I(A)$. They are the generic points of $V(I(A))$ (as a closed subset of the affine space) and also the generic points of A (as a closed subset of P^k).

From 4.2 follows:

Corollary 4.3 *Let $C \subseteq A$ be λ-closed subsets of P^k. Then C is an irreducible component of A (in P^k) iff $V(I(C))$ is an irreducible component of $V(I(A))$ (in the affine space).*

Proof: C is an irreducible component of A in P^k iff $I(C)$ is a type ideal containing $I(A)$ and minimal among ideals corresponding to a type completing P^k. But, in this case, $I(C)$ is also minimal among all type ideals containing $I(A)$. Indeed for Q such an ideal, $I(A) \subseteq Q \subseteq I(C)$, hence, for each $i = 1, \ldots, k$,

$$I(P_i) = I(A) \cap K[X_{i\infty}] \subseteq Q \cap K[X_{i\infty}] \subseteq I(C) \cap K[X_{i\infty}] = I(P_i),$$

hence $I(Q) \cap K[X_{i\infty}] = I(P_i)$. □

Remark 4.4 *An ideal I of $K[X_{1\infty},\dots,X_{k\infty}]$, separable and intersecting each $K[X_{i\infty}]$ in $I(P_i)$, is the canonical ideal of some λ-closed subset of the affine space, thus $I = \cap\mathcal{Q}$ for $\mathcal{Q} :=$ { minimal prime ideals containing I }. Let us define*

$$\mathcal{Q}_1 := \{Q \in \mathcal{Q}; Q \cap K[X_{i\infty}] = I(P_i),\ for\ i = 1,\dots,k\}.$$

Then $V(I) \cap P^k \neq \emptyset$ iff $\mathcal{Q}_1 \neq \emptyset$, and in this case $I(V(I) \cap P^k) = \cap\mathcal{Q}_1$.

Proof: 1. Clearly any $Q \in \mathcal{Q}_1$ is an $I(R)$ for some type R satisfying $R \subseteq V(I) \cap P^k$. Hence $\mathcal{Q}_1 \neq \emptyset$ implies that $V(I) \cap P^k \neq \emptyset$.
2. One has

$$V(I) \cap P^k \subseteq V(I) \cap V(I(P^k)),$$

hence by 3.7

$$I(V(I) \cap P^k) \supseteq I(V(I) \cap V(I(P^k))) = I + I(P^k) = I.$$

Consequently any $m \in V(I) \cap P^k$ satisfies $I(m) \supseteq I$ and, by definition of $\mathcal{Q}$, there is $Q \in \mathcal{Q}$ satisfying $I(m) \supseteq Q \supseteq I$. But, as we saw in the proof of 4.3, such a Q must belong to $\mathcal{Q}_1$. Thus $V(I) \cap P^k \neq \emptyset$ implies that $\mathcal{Q}_1 \neq \emptyset$.
3. Clearly $I(V(I) \cap P^k) \subseteq \cap\mathcal{Q}_1$. Conversely, let $f \in K[X_{1\infty},\dots,X_{k\infty}] \setminus I(V(I) \cap P^k)$ and $m \in V(I) \cap P^k$ satisfying $f(m) \neq 0$. As previously, there is $Q \in \mathcal{Q}$, hence $Q \in \mathcal{Q}_1$, satisfying $I \subseteq Q \subseteq I(m)$ and a fortiori $f \notin Q$. □

We can give an example of a separable ideal I intersecting each $K[X_{i\infty}]$ in $I(P_i)$ and not being the ideal of any λ-closed subset of P^k. Take $I = Q_1 \cap Q_2$ in $K[X_\infty, Y_\infty]$, for
$Q_1 := (I(P(X)), X - Y)$
$Q_2 := (X - a, Y - b)$,
where a and b are two distinct zeros of $I(P)$ in K_0. In this case $I \subset Q_1 = I(V(I) \cap P^2)$.

5 λ-closed subsets of a minimal type

We now add the further condition that the complete type we are working with is minimal (see definition after 2.15). We still have $K_0 \preceq L \models T_{p,\nu}$ and L is $|K_0|^+$-saturated; $P \in S_1(K_0)$ is a minimal type.

We are going to show that P is Zariski, in the sense of [Mar] section 5. First let us recall the meaning of the U-rank, (or Lascar rank) which is used in this definition, in our specific context.

Definitions: Suppose P minimal. Let $K_0 \preceq K \preceq L$ and $x_1,\dots,x_{k+1} \in P(L)$.

1. The rank over K of a tuple from P is defined inductively as follows:
   $rk(x_1; K) = 0$ if $x_1 \in K$
   $\qquad\qquad\ \ = 1$ *if not*,

and,

$$\begin{aligned} rk(x_1,...,x_k,x_{k+1};K) &= rk(x_1,...,x_k;K) && \text{if } x_{k+1} \in K\langle x_1,...,x_k\rangle \\ &= rk(x_1,...,x_k;K)+1 && \text{if not} \end{aligned}$$

(the consistancy of this definition follows from the minimality of P).

2. We will say that $x_1, \ldots, x_k$ are independent over K if $rk(x_1,...,x_k;K) = k$, or equivalently $I(x_1,...,x_k,K) = \sum I(P_i)$.

3. For A an $\wedge$-definable subset of P^k, with parameters from K_A, where $K_0 \preceq K_A \preceq L$ and $|K_0| = |K_A|$,

$$rk(A) := \max\{rk(x;K_A); x \in A(L)\}.$$

This rank does not depend on the choice of such a K_A (nor on L).

As particular $\wedge$-definable subsets A of P^k with few parameters, we will consider λ-closed subsets, with arbitrary parameters from L. Indeed as noted previously such an A requires only countably many parameters: the elements of a field of definition F_A of $I(A, L)$. We will consider A to be defined over some K_A, $F_A \subseteq K_A$, $K_0 \preceq K_A \preceq L$, $|K_A| = |K_0|$, to follow the conventions of the previous sections. We will say that A is irreducible if it is irreducible as a λ-closed set over K_A. The following facts tell us that working over such a field of definition is legitimate.

Fact 5.1 1) *Let $K \preceq L$ and I be an ideal of $K[X_{1\infty}, \ldots, X_{k\infty}]$. Then I is prime or separable iff $I \otimes_K L$ is.*
2) *Let A be a λ-closed subset of L^k defined over K_A, $K_A \preceq K \preceq L$. Then A is irreducible as a K_A-closed set iff it is irreducible as a K-closed set.*

Proof: 1) Let J be an ideal of some ring $K[X_1, \ldots, K_m]$, generated by $f_1, \ldots, f_n$ and for each $i \leq n$, let b_i denote the sequence of coefficients of f_i. Then the fact that J is prime, or separable, is a first order property of the $b_i's$ in K (see for example [De 88]). The result then follows by elementary inclusion.
2) is a direct consequence of 1). □

Fact 5.2 *Let A be an irreducible λ-closed subset of $P(L)^k$, defined over K_A, satisfying $K_0 \preceq K_A \preceq L$ and $|K_0| = |K_A|$. Then $P(L)^k$ contains generic points over K_A and one has for such a's,*

$$rk(A) = rk(a;K_A).$$

Proof: The existence of such a's follows from the $|K_0|^+$-saturation of L. Now $a \in P(L)^k$ is generic over K_A iff $I(a, K_A) = I(A, K_A) \subseteq I(x, K_A)$, for each $x \in A$. If $rk(x;K_A) = r = rk(x_{i_1},...,x_{i_r};K_A)$, we have $I(x,K_A) \cap K_A[X_{i_1},...,X_{i_r}] = \sum_{j=1}^r I(P_{i_j})$, which implies $I(a,K_A) \cap K_A[X_{i_1},...,X_{i_r}] = \sum_{j=1}^r I(P_{i_j})$ and hence that $rk(a;K_A) \geq r$. □

We will see below (Proposition 5.6) that any λ-closed subset of a minimal type is of finite type. Hence in order to show that any minimal type is a Zariski geometry in the sense of [Mar] section 5, we must show that the following conditions hold.

i) Every λ-closed set in P^k is a finite union of irreducible λ-closed sets.

ii) If $A \subset B$ are λ-closed subsets of P^k and B is irreducible, then $rk(A) < rk(B)$.

iii) (Dimension theorem) If A is a λ-closed irreducible subset of P^k, $rk(A) = m$ and $B = \{\overline{x} \in P^k : x_i = x_j\}$, then $rk(C) \geq m-1$ for every non-empty irreducible component C of $A \cap B$.

Proposition 5.3 *Let P be minimal, $(a_1, ..., a_k) \in P^k$ and K, $K_0 \preceq K \preceq L$. Suppose that $a_1, ..., a_r$ are independent over K, and that $a_{r+1}, ..., a_k$ are algebraic over $(K, a_1, ..., a_r)$, hence separably algebraic over $K(a_{1\infty}, ..., a_{r\infty})$. Choose n such that*

$$[K(a_{1\infty}, ..., a_{r\infty})(a_{r+1}, ..., a_k) : K(a_{1\infty}, ..., a_{r\infty})] =$$

$$[K(a_{1\leq n}, ..., a_{r\leq n}, a_{r+1}, ..., a_k) : K(a_{1\leq n}, ..., a_{r\leq n})].$$

Let $Q = I(a_1, ..., a_k, K)$ and define, for $m \in \omega$,

$$M_m = K[X_{1\leq n+m}, ..., X_{r\leq n+m}, X_{r+1\leq m}, ..., X_{k\leq m}].$$

(1) There is $d \in K[X_{1\leq n}, ..., X_{r\leq n}] \setminus \sum_{i=1}^{r} I(P_i)_{\leq n}$ such that, for each $i = r+1, ..., k$, $m \in \mathbb{N}$ and $j \in p^{\nu m}$,

$$a_i \in K[a_{1\leq n}, ..., a_{r\leq n}, a_i^{p^m}, d(a_1, ..., a_r)^{-1}],$$

$$a_{i,j} \in K[a_{1\leq n+m}, ..., a_{r\leq n+m}, a_i, d(a_1, ..., a_r)^{-1}].$$

(2) $h(P_1 \otimes ... \otimes P_r; K)(x_1, ..., x_r) \cup \{f_1(x) = ... = f_l(x) = 0\} \vdash tp(a_1, ..., a_k; K)$, for $f_1, ... f_l$ generating $Q \cap M_0$.

(3) Q is the unique prime separable ideal of $K[X_{1\infty}, \ldots, X_{k\infty}]$ which intersects $K[X_{1\infty}, ..., X_{r\infty}]$ in $\sum_{i=1}^{r} I(P_i)$ and contains $\sum_{i=r+1}^{k} I_i^0$ and $Q \cap M_0$.

(4) $Q \cap M_m$ is the unique prime separable ideal of M_m which intersects $K[X_{1\leq n+m}, ..., X_{r\leq n+m}]$ in $\sum_{i=1}^{r} I(P_i)_{\leq n+m}$, and contains $\sum_{i=r+1}^{k} I_{i\leq m}^0$ and $Q \cap M_0$.

(5) $Q \cap M_m$ is the unique minimal prime separable ideal of M_m which intersects each $K[X_{i\leq n+m}]$ in $I(P_i)_{\leq n+m}$ for $i = 1, ..., r$, and each $K[X_{i\leq m}]$ in $I_{i\leq m}^0$ for $i = r+1, ..., k$, contains $Q \cap M_0$ and not d.

(6) Q is the unique minimal prime separable ideal of $K[X_{1\infty}, \ldots, X_{k\infty}]$ which intersects each $K[X_{i\infty}]$ in $I(P_i)$ for $i = 1, ..., k$, contains $Q \cap M_0$ and not d.

Proof: (1) Since each a_i is separable over $K(a_{1\leq n}, ..., a_{r\leq n})$, by 2.5, there is a polynomial $d \in K[X_{1\leq n}, ..., X_{r\leq n}] \setminus \sum_{i=1}^{r} I(P_i)_{\leq n}$ such that, for $i = r+1, ..., k$ and for all $m \in \mathbb{N}$,

$$a_i \in K[a_{1\leq n}, ..., a_{r\leq n}, a_i^{p^m}, d(a_1, ..., a_r)^{-1}].$$

Hence, for some $e \in \mathbb{N}$,

$$d(a_1, ..., a_r)^{p^m.e}.a_i = b_i(a_1, ..., a_r, a_i)$$

with $b_i \in K[X_{1\leq n}, ..., X_{r\leq n}, X_i^{p^m}]$. We can decompose

$$b_i(X_1, ..., X_r, X_i) \equiv \sum_{j\in p^{\nu m}} b_{ij}(X_1, ..., X_r, X_i)^{p^m}.m_j \text{ (modulo } \sum_{i=1}^{r} I_i^0)$$

with $b_{ij} \in K[X_{1\leq n+m}, ..., X_{r\leq n+m}, X_i]$, which proves

$$a_{i,j} = b_{ij}(a_1, ..., a_r, a_i).d(a_1, ..., a_r)^{-e} \in K[a_{1\leq n+m}, ..., a_{r\leq n+m}, a_i, d(a_1, ..., a_r)^{-1}].$$

(2) is an equivalent formulation of (3).

(3) If Q' is an ideal of type satisfying the conditions, it does not contain d and it contains the polynomials

$$d(X_1, ..., X_r)^{p^m.e}.X_i - b_i(X_1, ..., X_r, X_i)$$

for $i = r+1, ..., k$, because these polynomials belong to $M_0 \cap Q$. Since Q' contains $\sum_1^k I_i^0$ and is separable, the proof of (1) above shows that Q' contains the polynomials

$$d(X_1, ..., X_r)^e.X_{ij} - b_{ij}(X_1, ..., X_r, X_i).$$

Hence Q' describes the same type as Q, and $Q' = Q$.

(4) For q such an ideal of M_m, M_m/q is K-isomorphic to $K[a_{1\leq n+m}, ..., a_{r\leq n+m}, a_{r+1\leq m}, ..., a_{k\leq m}]$.

(5) Let q be an ideal satisfying the condition. By the proof of (3), it contains $d(X_1, ..., X_r)^e.X_{ij} - b_{ij}(X_1, ..., X_r, X_i)$. Let $R_m = M_m[d^{-1}]$. In R_m, the ideal generated by $Q \cap M_0, \sum_{i=1}^{k}(I(P_i) \cap M_m)$, and the polynomials $d(X_1, ..., X_r)^e.X_{ij} - b_{ij}(X_1, ..., X_r, X_i)$ for $i = r+1, ..., k$ and $j \in p^{\nu m}$, is prime and therefore equals the ideal generated by $Q \cap M_m$. Hence $R_m.q \supseteq R_m.(Q \cap M_m)$. Since $d \notin q$, $R_m.q$ and $R_m.(Q \cap M_m)$ intersect M_m respectively in q and $Q \cap M_m$, therefore $q \supseteq Q \cap M_m$.

(6) is obvious from (5). □

Proposition 5.4 *Let P be minimal and $A \subset B$ be two λ-closed irreducible subsets of P^k. Then $rk(A) < rk(B)$.*

Proof: We assume A non empty. Define $r := rk(A), l := rk(B)$. By definition of the rank, $r \leq l$. Let K be a field containing fields of definition for $I(A, L)$ and $I(B, L)$, chosen such that $K_0 \preceq K \preceq L$ and $|K_0| = |K|$. In the rest of the proof, genericity, independance and canonical ideals are relative to K. Let a and b be generics of A and B respectively, then $I(a) = I(A) \supseteq I(b) = I(B)$. Hence, if $a_{i_1}, ..., a_{i_r}$ are independent, so are $b_{i_1}, ..., b_{i_r}$. Reorder the indices so that $a_1, ..., a_r$ are independent, and $b_1, ..., b_l$ are independent. Choose n so that $b_{l+1}, ..., b_k$ are separably algebraic over $K(b_{1\leq n}, ..., b_{l\leq n})$ and $a_{r+1}, ..., a_k$ separably algebraic over $K(a_{1\leq n}, ..., a_{r\leq n})$, and define

$$M_m = K[X_{1\leq n+m}, ..., X_{l\leq n+m}, X_{l+1\leq m}, ..., X_{k\leq m}].$$

Then $\dim(I(B) \cap M_m) = \dim(I(B) \cap K[X_{1\leq n+m}, ..., X_{l\leq n+m}])$ and

$$\dim(I(A) \cap M_m) = \dim(I(A) \cap K[X_{1\leq n+m}, ..., X_{l\leq n+m}]),$$

where "dim" here denotes the Krull dimension of the ideal. Now,

$$I(A) = \cup_m (I(A) \cap M_m),$$

$$I(B) = \cup_m (I(B) \cap M_m),$$

and $I(A) \supset I(B)$. So for some m, $I(A) \cap M_m$ strictly contains $I(B) \cap M_m$. Since they are both prime, this implies $\dim(I(A) \cap M_m) < \dim(I(B) \cap M_m)$, and therefore $I(A) \cap K[X_{1\leq n+m}, ..., X_{l\leq n+m}]$ strictly contains $I(B) \cap K[X_{1\leq n+m}, ..., X_{l\leq n+m}]$, i.e. $a_1, ..., a_l$ are not independent. □

Corollary 5.5 *Idem with A non necessarily irreducible.*

Proof: A is the union of its irreducible λ-components. □

Proposition 5.6 *If P is minimal, any λ-closed subset of P^k is of finite type.*

Proof: The proof is by induction on the rank of a λ-closed subset A of P^k. If $rk(A) = 0$ then (by compactness and by $|K_0|^+$-saturation of L) A is finite. Suppose the result true for all λ-closed subsets of P^k of rank $< r$.

Claim: If R is a complete type of rank $\leq r$ realized in P^k, then $V(I(R) \cap P^k)$ is of finite type.

Proof of the Claim: Take a field K, containing a field of definition for $I(R, L)$ and satisfying $K_0 \preceq K \preceq L$ and $|K_0| = |K|$, and consider ranks and canonical ideals above K. As in 5.3, find n and d and define M_m such that $I(R)$ is the unique minimal prime separable ideal of $K[X_{1\infty}, \ldots, X_{k\infty}]$ intersecting each $K[X_{i\infty}]$ in $I(P_i)$, containing $I(R) \cap M_0$ and not d. By 5.5, as $I(R)$ is irreducible, the λ-closed subset B of P^k defined by $B := V(I(R), d) \cap P^k$ has rank $\leq r - 1$, and therefore, by induction hypothesis,

$$B = V(I(R) \cap M_m, d) \cap P^k,$$

for some m. We prove now that $V(I(R)) \cap P^k = V(I(R) \cap M_m) \cap P^k$. Let $a \in V(I(R) \cap M_m) \cap P^k$. If $d(a) \neq 0$ then $I(a)$ contains $I(R)$, as it contains $\sum_{i=1}^k I(P_i)$, $I(R) \cap M_0$ and not d. If $d(a) = 0$, then $a \in B$, hence $a \in V(I(R))$. □

Let now A be a λ-closed subset of P^k of rank r, defined over K, $K_0 \preceq K \preceq L$ and $|K_0| = |K|$, and R_α, $\alpha < \alpha_0$, the distinct complete k-types over K completing A. Hence, working over K, we have the equivalence

$$P^k(x) \vdash x \in A \leftrightarrow \vee_{\alpha<\alpha_0} R_\alpha(x) \leftrightarrow \vee_{\alpha<\alpha_0} x \in V(I(R_\alpha)).$$

Now, each R_α has rank $\leq r$ and, by the claim, modulo P^k, each $V(I(R_\alpha))$ is of finite type. Hence the above equivalence holds between an infinite conjunction (the definition of A) and an infinite disjunction. By compactness it is also equivalent to a finite subdisjunction. □

Proposition 5.7 *If P is minimal, each λ-closed subset of P^k is union of finitely many irreducible λ-closed subsets of P^k.*

Proof: Follows from the proof of 5.6. □

Now we can prove the Dimension theorem:

Proposition 5.8 *Let P be minimal, A an irreducible λ-closed subset of P^k of rank r, B a diagonal hyperplane of equation $X_{i_1} = X_{i_2}$, for some $i_1, i_2 \in \{1, ..., k\}$, and C a non-empty irreducible component of $A \cap B$. Then $rk(C) \geq r - 1$.*

The proof is rather long. The key argument is given by the following claim:

Claim: Let I and J be ideals of non empty λ-closed subsets of P^k, Q a minimal prime ideal containing $I + J$ and K, containing a field of definition of I, J and Q satisfying $K_0 \preceq K \preceq L$ and $|K_0| = |K|$. Suppose further that a is a generic zero of Q over K, $a \in P^k$, and that

$$rk(V(Q) \cap P^k) = rk(a_1, ..., a_l, K) = l.$$

Let $n \in \mathbb{N}$ be such that $a_{l+1}, ..., a_k$ are separably algebraic over $K(a_{1\leq n}, ..., a_{l\leq n})$, with

$$[K(a_{1\infty}, ..., a_{l\infty}, a_{l+1}, ..., a_k) : K(a_{1\infty}, ..., a_{l\infty})] =$$
$$[K(a_{1\leq n}, ..., a_{l\leq n}, a_{l+1}, ..., a_k) : K(a_{1\leq n}, ..., a_{l\leq n})].$$

For $m, n_1, ..., n_k \in \mathbb{N}$, define

$$M_{m,(n_1,...,n_k)} := K[X_{1\leq n_1+m}, ..., X_{k\leq n_k+m}],$$

and for any $\mathcal{I} \subseteq K[X_{1\infty}, \ldots, X_{k\infty}]$, $\mathcal{I}_{m,(n_1,...,n_k)} := M_{m,(n_1,...,n_k)} \cap \mathcal{I}$.
Then there exists some integer N such that for each $m, n_1, ..., n_k$ satisfying $n_1, ..., n_l \geq n + n_{l+1}, ..., n + n_k$, $Q_{m,(n_1,...,n_k)}$ is the intersection of $M_{m,(n_1,...,n_k)}$ with some minimal prime ideal of $M_{N+m,(n_1,...,n_k)}$ containing $I_{N+m,(n_1,...,n_k)} + J_{N+m,(n_1,...,n_k)}$.

Note that, for $n_1 = ... = n_l = n$, $n_{l+1} = ... = n_k = 0$ and $\overline{n} := (n_1, ..., n_k)$, the ring considered in and after 5.3,

$$M_m := K[X_{1\leq n+m}, ..., X_{l\leq n+m}, X_{l+1\leq m}, ..., X_{k\leq m}],$$

is equal to $M_{m,\overline{n}}$. Also a particular case of the claim asserts that $Q \cap M_m$ is the intersection of M_m with some ideal of M_{N+m} which is minimal prime over $I_{N+m} + J_{N+m}$.

Proof of the Claim: In order to simplify notation, we rename
$Y = (X_{1\leq n}, ..., X_{l\leq n})$
X any of the variables $X_{l+1}, ..., X_k$, say X_{l+1}.
by Proposition 5.3 there is $d \in K[Y] \setminus Q$ such that

$$a_{l+1} \in K[a_{1\leq n}, ..., a_{l\leq n}, d(a_1, ..., a_l)^{-1}, a_{l+1}^{p^m}],$$

for each integer m. Hence, for each m, there are some $e \in K[X, Y]$ and $\sigma \in \mathbb{N}$, such that

$$(*) \qquad d(Y)^{p^m.\sigma}.X + e(X^{p^m}, Y) \in Q \cap M_0.$$

Now $I + J$ is separable by 3.8 and by the Nullstellensatz is an intersection of type ideals, which we can choose minimal. By 5.7 only finitely many of them, say $Q_1, ...Q_r$, correspond to types completing P^k, hence

$$I + J = Q_1 \cap ... \cap Q_r \cap S$$

with $V(S) \cap P^k = \emptyset$. By 4.3 each Q_i is minimal over $I + J$ and by 3.10, Q is one of them, say Q_1. Take $g \in Q_2 \cap ... \cap Q_r \cap S \setminus Q$. Then $g.Q \subseteq I + J$. Fix $N \in \mathbb{N}$ such that

$$g \in K[X_{\leq N}, Y_{\leq N}], \text{ and further}$$

$$g.(Q \cap M_0) \subseteq (I \cap M_N) + (J \cap M_N).$$

Then, by $(*)$, there exist $u \in I \cap M_N$ and $v \in J \cap M_N$ such that

$$g^{p^m}.[d(Y)^{p^m.\sigma}.X + e(X^{p^m}, Y)] = u + v.$$

Let us decompose

$$e(X^{p^m}, Y) \equiv \sum_{j \in p^{\nu m}} e_j^{p^m}.m_j \text{ (modulo } (\sum_i I_i^0) \cap M_m)$$

$$u \equiv \sum_{j \in p^{\nu m}} u_j^{p^m}.m_j \text{ (modulo } (\sum_i I_i^0) \cap M_{N+m})$$

$$v \equiv \sum_{j \in p^{\nu m}} v_j^{p^m}.m_j \text{ (modulo } (\sum_i I_i^0) \cap M_{N+m}),$$

where $e_j \in K[X, Y_{\leq m}], u_j$ and $v_j \in M_{N+m}$ and more precisely, since I and J are separable, $u_j \in I \cap M_{N+m}$ and $v_j \in J \cap M_{N+m}$. Therefore

$$\sum [g.(d^\sigma.X_j + e_j) - u_j - v_j]^{p^m}.m_j \in \sum_i I_i^0 \cap M_{N+m},$$

hence, by separability of $\sum_i I_i^0$,

$$g.(d^\sigma.X_j + e_j) - u_j - v_j \in \sum_i I_i^0$$

and finally, since I and J contain $\sum_i I_i^0$,

$$g.(d^\sigma.X_j + e_j) \in (I \cap M_{N+m}) + (J \cap M_{N+m}).$$

We argue now for $X_{l+2}, ..., X_k$ just as we did for $X = X_{l+1}$, and so get an integer N and polynomials $\alpha_i \in K[X_{1\leq n}, ..., X_{l\leq n}] \setminus Q$ and $g_i \in K[X_{i\leq N}, X_{1\leq n+N}, \ldots, X_{l\leq n+N}] \setminus Q$, for $i = l+1, ...k$, such that:

for all $m \in \mathbb{N}$ and $j \in p^{\nu m}$, there are $\beta_{i,j} \in K[X_i, X_{1\leq n+N+m}, ..., X_{l\leq n+N+m}]$ for which $g_i.(\alpha_i.X_{i,j} + \beta_{i,j}) \in (I \cap M_{N+m}) + (J \cap M_{N+m})$.

By applying this result to $j \in p^{\nu(m+n_i)}$, we get

$$g_i.(\alpha_i.X_{i,j} + \beta_{i,j}) \in (I \cap M' + J \cap M'),$$

with

$$M' := K[X_i, X_{1\leq n+N+m+n_i}, ..., X_{l\leq n+N+m+n_i}]$$
$$\subseteq K[X_i, X_{1\leq N+m+n_1}, ..., X_{l\leq N+m+n_l}] \subseteq M_{N+m,\overline{n}}.$$

Since there is at most one prime ideal of $M_{m,\overline{n}}$ intersecting $K[X_{1\infty}, ..., X_{l\infty}]$ in $\sum_1^l I(P_i)_{\leq n_i+m}$, containing $Q \cap M_0$, each $g_i.(\alpha_i.X_{i,j} + \beta_{i,j})$ and no $\alpha_i.g_i$ for $i = l+1, ..., k$ and $j \in p^{\nu(m+n_i)}$, every prime ideal of $M_{N+m,\overline{n}}$ intersecting $K[X_{1\infty}, ..., X_{l\infty}]$ in $\sum_1^l I(P_i)_{\leq n_i+m+N}$, containing $I_{N+m,\overline{n}}$ and $J_{N+m,\overline{n}}$ and no $\alpha_i.g_i$, intersects $M_{m,\overline{n}}$ in $Q_{m,\overline{n}}$. □

Proof of Proposition 5.8 : Let us choose once more a field K, containing fields of definition for $I(A, L), I(B, L)$ and $I(C, L)$, satisfying $K_0 \preceq K \preceq L$ and $|K_0| = |K|$. Independence and genericity are considered relatively to this K.

Definition: We will say that $X_{i_1}, ..., X_{i_l}$ are <u>independent modulo some irreducible λ-closed subset F of P^k</u> if , for $(a_1, ..., a_k)$ a generic point of F,

$a_{i_1}, ..., a_{i_l}$ are independent. One variable $X_{i_{l+1}}$ is algebraic over $X_{i_1}, ..., X_{i_l}$ modulo F if $a_{i_{l+1}} \in K\langle a_{i_1}, ... a_{i_l}\rangle$.

We come back to the proof of 5.8. If $A \cap B = \emptyset$ or $A \subseteq B$, there is nothing to prove. Suppose therefore that $A \cap B$ is non empty and that $A \cap B \neq A$.

I. We consider first the case where X_{i_1} and X_{i_2} are independent modulo A and X_{i_1} independent modulo C. Thus we are allowed to rename $X_1, ..., X_k$ as X, U, V, Y, Z where
- X, Y, Z are tuples, we will then denote their length by ℓ,
- U and V are variables,
- modulo A, (X, U, V, Y) is independent, and Z algebraic over (X, U, V, Y),
- the equation defining B is $U = V$,
- modulo C, (X, U) is independent, and V, Y, Z are algebraic over (X, U)
(U and V are given, choose X and then Y). So we have to prove that there is in fact no Y, in other words that $\ell Y = 0$.

Let $n \in \mathbb{N}$ be such that
- modulo $I(A)$, Z is algebraic over $K(X_{\leq n}, U_{\leq n}, V_{\leq n}, Y_{\leq n})$ of same degree as over $K(X_\infty, U_\infty, V_\infty, Y_\infty)$,
- modulo $I(C)$, V, Y and Z are algebraic over $K(X_{\leq n}, U_{\leq n})$ of same degree as over $K(X_\infty, U_\infty)$.
We are going to work with variables $X_{\leq 2n+m+m'}, U_{\leq 2n+m+m'}, V_{\leq n+m}, Y_{\leq n+m+m'}$ and $Z_{\leq m}$, where m and m' are arbitrary integers. Note that the depth of the components is chosen in order to witness the dependence relation modulo A or C. The corresponding polynomial ring is

$$M_{m,m'} := K[X_{\leq 2n+m+m'}, U_{\leq 2n+m+m'}, V_{\leq n+m}, Y_{\leq n+m+m'}, Z_{\leq m}].$$

For an ideal $\mathcal{I}$ of $K[X_{1\infty}, \ldots, X_{k\infty}]$ we define

$$\mathcal{I}_{m,m'} := \mathcal{I} \cap M_{m,m'}.$$

By Lemma 2.6, the ideals $I(A)_{m,m'}, I(B)_{m,m'}, I(C)_{m,m'}$ and, for all integer s, $I(P)_{\leq s}$ are absolutely prime and therefore define over L^a irreducible varieties, which we will denote by $A_{m,m'}, B_{m,m'}, C_{m,m'}$ and $\overline{P}_{\leq s}$. The varieties $A_{m,m'}, B_{m,m'}$ and $C_{m,m'}$ are subvarieties of

$$E_{m,m'} := \overline{P}^{\ell X+1}_{\leq 2n+m+m'} \times \overline{P}_{\leq n+m} \times \overline{P}^{\ell Y}_{\leq n+m+m'} \times \overline{P}^{\ell Z}_{\leq m}.$$

Let us now compute the dimensions of all these varieties. A generic point of $A_{m,m'}$ is the projection of some generic point of A (identifying x and x_∞ as in 3.12), and the same holds for $B_{m,m'}, C_{m,m'}, \overline{P}_{\leq s}$ and $E_{m,m'}$. Hence we have, using the function $f(s) := \dim(\overline{P}_{\leq s})$,

$\dim(A_{m,m'}) = (\ell X + 1).f(2n + m + m') + f(n + m) + \ell Y.f(n + m + m')$

$\dim(B_{m,m'}) = (\ell X + 1).f(2n + m + m') + \ell Y.f(n + m + m') + \ell Z.f(m)$

$\dim(C_{m,m'}) = (\ell X + 1).f(2n + m + m')$

$$\dim(E_{m,m'}) = (\ell X+1).f(2n+m+m')+f(n+m)+\ell Y.f(n+m+m')+\ell Z.f(m).$$

By the claim, $C_{m,m'}$ is the projection over $E_{m,m'}$ of some irreducible component D of $A_{m+N,m'} \cap B_{m+N,m'}$ where N is independent of m and m'. This implies, if $\pi : E_{m+N,m'} \to E_{m,m'}$ is the canonical projection,

$$\dim(C_{m,m'}) = \dim D - \dim(\pi^{-1}(c))$$

for c a generic of $C_{m,m'}$, hence

$$\dim(C_{m,m'}) \geq \dim D + \dim(E_{m,m'}) - \dim(E_{m+N,m'}).$$

Consider the set

$$E := P^{\ell X+1}_{\leq 2n+m+m'+N} \times P_{\leq n+m+N} \times P^{\ell Y}_{\leq n+m+m'+N} \times P^{\ell Z}_{\leq m+N}$$

As $C \neq \emptyset, D \cap E \neq \emptyset$ too. Furthermore each point of $D \cap E$ is simple on D as each of its coordinates is generic, hence simple in the corresponding irreducible variety (see [Lan 58, VIII section 2, Prop.6]:

- $x_{1 \leq 2n+m+m'+N}, \ldots, x_{\ell X+1 \leq 2n+m+m'+N}$ are each generic points of $\overline{P}_{\leq 2n+m+m'+N}$,
- $x_{\ell X+2 \leq n+m+N}$ is generic in $\overline{P}_{\leq n+m+N}$,
- $x_{\ell X+3 \leq n+m+m'+N}, \ldots, x_{\ell X+\ell Y+2 \leq n+m+m'+N}$ are each generics of $\overline{P}_{\leq n+m+m'+N}$,
- $x_{\ell X+\ell Y+3 \leq m+N}, \ldots, x_{k \leq m+N}$ are each generics of $\overline{P}_{\leq m+N}$.

We can then apply, inside $E_{m+N,m'}$, the classical dimension theorem of algebraic geometry (see for instance [Lan 58, VIII section 5, theorem 5]), hence

$$\dim D \geq \dim(A_{m+N,m'}) + \dim(B_{m+N,m'}) - \dim(E_{m+N,m'}),$$

therefore

$$\dim(C_{m,m'}) \geq \dim(A_{m+N,m'}) + \dim(B_{m+N,m'}) + \dim(E_{m,m'}) - 2\dim(E_{m+N,m'}).$$

Substituting in this we obtain

$$0 \geq f(n+m) + \ell Y.f(n+m+m') + \ell Z.f(m) - \ell Z.f(m+N) - f(n+m+N).$$

On the second side of this inequality, m' occurs only in the second term, and we can choose it arbitrarily big. How do all these terms behave when one of m and m' goes to the infinity ?

1. If f is bounded on $\mathbb{N}$ (P is then said to be <u>thin</u>, thin types play a main role in the proof of Hrushovski, see the next section), all terms $f(m + ...)$ become equal for m big enough, which forces $\ell Y = 0$.
2. If $f(s) \to \infty$ for $s \to \infty$, let us fix m and let m' tend to ∞. Again ℓY must equal 0.

II. We consider now the case where X_{i_1} is independent modulo A, but (X_{i_1}, X_{i_2}) is not. Let $X \subseteq \{X_1, ..., X_k\}$ be a maximal independent subset modulo C.

Claim: Then X_{i_1} and X_{i_2} are algebraic over C, and either $X\frown X_{i_1}$ or $X\frown X_{i_2}$ is independent modulo A.

Proof: From $A \not\subseteq B$ follows $C \subset A$, therefore $I(A) \subset I(C)$, but also

$$(**) \qquad I(A) \cap K[X_\infty, X_{i_1,\infty}, X_{i_2,\infty}] \subset I(C) \cap K[X_\infty, X_{i_1,\infty}, X_{i_2,\infty}]$$

(if one of the variables X_{i_1} or X_{i_2} belongs to X, we do not repeat it in the polynomial ring above). Indeed, since the polynomials $X_{i_1,j} - X_{i_2,j}, j \in p^\infty$, all are in $I(C) \cap K[X_\infty, X_{i_1,\infty}, X_{i_2,\infty}]$ and, together with $\sum_{i=1}^k I_i^0$, generate $I(B)$, the equality

$$I(A) \cap K[X_\infty, X_{i_1,\infty}, X_{i_2,\infty}] = I(C) \cap K[X_\infty, X_{i_1,\infty}, X_{i_2,\infty}]$$

would imply $I(A) \supseteq I(B)$, whence $A \subseteq B$. Now X is independent modulo C and X_{i_1} and X_{i_2} are dependent modulo A. Then the conclusion of the claim follows from the strict inclusion (**). □

Thus we are allowed as in the first case to reorder and rename $X_1, ..., X_k$ as X, U, V, Y, Z (X, Y, Z are tuples, U and V variables) such that
- X is independent maximal modulo C,
- (X, U, Y) is independent maximal modulo A, and
- the equation defining B is $U = V$.

We choose $n \in \mathbb{N}$ such that
- modulo $I(A)$, Z and V are algebraic over $K(X_{\leq n}, U_{\leq n}, Y_{\leq n})$ of same degree as over $K(X_\infty, U_\infty, Y_\infty)$,
- modulo $I(C)$, U, V, Y and Z are algebraic over $K(X_{\leq n})$ of same degree as over $K(X_\infty)$.

We define

$$M_{m,m'} := K[X_{\leq 2n+m+m'}, U_{\leq n+m+m'}, V_{\leq m}, Y_{\leq n+m+m'}, Z_{\leq m}],$$

and the varieties naturally associated $A_{m,m'}, B_{m,m'}, C_{m,m'}$ and $E_{m,m'}$, and analogously to the first case, the integer N and the subvariety D of $E_{m+N,m'}$. Then

$\dim(A_{m,m'}) = \ell X.f(2n+m+m') + (\ell Y + 1).f(n+m+m')$

$\dim(B_{m,m'}) = \ell X.f(2n+m+m') + (\ell Y + 1).f(n+m+m') + \ell Z.f(m)$

$\dim(C_{m,m'}) = \ell X.f(2n+m+m')$

$\dim(E_{m,m'}) = \ell X.f(2n+m+m') + (\ell Y + 1).f(n+m+m') + (\ell Z + 1).f(m).$

The necessary inequality becomes

$$0 \geq -(\ell Z + 2).f(m+N) + (\ell Y + 1).f(n+m+m') + (\ell Z + 1).f(m),$$

which forces P to be thin and then ℓY to be zero.

III. The case where both X_{i_1} and X_{i_2} are algebraic modulo A is prohibited since this would imply that $a \in A \Rightarrow a_{i_1} = a_{i_2}$, therefore $A \subseteq B$.

IV. There remains to be considered the case where X_{i_1} and X_{i_2} are independent modulo A and become both algebraic modulo C. We introduce $X \subseteq \{X_1, ..., X_k\}$ maximal independent modulo C and consider as in the second case

$$K[X_\infty, X_{i_1,\infty}, X_{i_2,\infty}]$$

(without repetition) and the trace over this ring of the strict inclusion $I(A) \subset I(C)$. The same argument proves that either $X \frown X_{i_1}$ or $X \frown X_{i_2}$ is independent modulo A.

Let us prove that $X \frown X_{i_1} \frown X_{i_2}$ can not be independent modulo A. In this case, define $U := X_{i_1}$, $V := X_{i_2}$, Y such that X, U, V, Y are maximal independent modulo A, and Z the other variables. We consider as previously $n \in \mathbb{N}$ and

$$M_{m,m'} := K[X_{\leq 2n+m+m'}, U_{\leq n+m+m'}, V_{\leq n+m}, Y_{\leq n+m+m'}, Z_{\leq m}],$$

the varieties $A_{m,m'}, B_{m,m'}, C_{m,m'}$ and $E_{m,m'}$, the integer N and the variety D. Then

$\dim(A_{m,m'}) = \ell X.f(2n+m+m') + (\ell Y+1).f(n+m+m') + f(n+m)$

$\dim(B_{m,m'}) = \ell X.f(2n+m+m') + (\ell Y+1).f(n+m+m') + \ell Z.f(m)$

$\dim(C_{m,m'}) = \ell X.f(2n+m+m')$

$\dim(E_{m,m'}) = \ell X.f(2n+m+m') + (\ell Y+1).f(n+m+m') + f(n+m) + \ell Z.f(m)$,

and the inequality becomes

$$0 \geq -\ell Z.f(m+N) - f(n+m+N) + (\ell Y+1).f(n+m+m') + f(n+m) + \ell Z.f(m),$$

which is seen first implying that P is thin and then being impossible.

So we are in the case where

- (X, U, Y) is maximal independent modulo A, and V, Z are the other variables, and
- X is maximal independent modulo C.

Taking

$$M_{m,m'} := K[X_{\leq 2n+m+m'}, U_{\leq n+m+m'}, Y_{\leq n+m+m'}, V_{\leq m}, Z_{\leq m}],$$

we get

$\dim(A_{m,m'}) = \ell X.f(2n+m+m') + (\ell Y+1).f(n+m+m')$

$\dim(B_{m,m'}) = \ell X.f(2n+m+m') + (\ell Y+1).f(n+m+m') + \ell Z.f(m)$

$\dim(C_{m,m'}) = \ell X.f(2n+m+m')$

$\dim(E_{m,m'}) = \ell X.f(2n+m+m') + (\ell Y+1).f(n+m+m') + (\ell Z+1).f(m)$, and the inequality

$$0 \geq -(\ell Z+2).f(m+N)+(\ell Y+1).f(n+m+m')+(\ell Z+1).f(m),$$

implies that P is thin and ℓY zero. □

Theorem 5.9 *Minimal types are Zariski.*

Proof: Condition i) is Proposition 5.7, ii) is Corollary 5.5 and iii) is Proposition 5.8. □

6 Thin types

Let $K \models T_{p,\nu}$.

Definition: A type $P \in S_1(K)$ is thin if the transcendence degree $tr(K\langle P\rangle, K)$ is finite. In this case, we define $RT(P) := tr(K\langle P\rangle, K)$. Otherwise $RT(P) = \infty$.

Proposition 6.1 1. *RT is a rank.*
2. *A thin type P is ranked and $RU(P) \leq RT(P)$.*

Proof: 1. follows from the characterization of forking we gave in section 2.
2. is proved by induction on $RT(P)$. For x realizing P and $L \succeq K$, $t(x, L)$ forks over K iff L and $K\langle x\rangle$ are not linearly disjoint over K iff $tr(L\langle x\rangle, L) < tr(K\langle x\rangle, K)$. □

Examples: 1. For $L \succeq K$ and P the type over K of an element of $L^{p^\infty} \setminus K$, $RU(P) = RT(P) = 1$.
2. If x and y are two independent realizations of the previous P and $z = x^p + by^p$ (b an element in the p-basis of K), then $RU(x, K) = RT(x, K) = 2$.
3. In [CCSSW], some types with $RU = 1$ and $RT = 2$ are described, and in [CWo], some minimal non thin types are constructed.

Thin types arise naturally in the context of algebraic groups over separably closed fields as was shown in [Hr 96, Lemma 2.15]:

Proposition 6.2 *Let $\underline{G}$ be an Abelian algebraic group defined over $K \models T_{p,\nu}$, $G := \underline{G}(K)$, $A := \cap p^n G$. Then generic types of A are thin.*

Claim (Weil, [We 48]): Let U be a variety defined over a field F of characteristic p, $M \in U$ generic over F and f a rational map $f : U^{p^n} \dots \to F$, where n is an arbitrary integer, f defined with coefficients from F, defined at $(M, \dots, M) \in U^{p^n}$

and symmetrical, which means that $f(M_1, ..., M_{p^n}) = f(M_{\sigma(1)}, ..., M_{\sigma(p^n)})$ for any permutation σ of $\{1, ..., p^n\}$. Then there is a rational map $g : U... \to F$ defined over $F^{p^{-n}}$, defined at M and such that $f(M, ..., M) = g(M)^{p^n}$.

The proof of this claim uses derivations, see [We 48, I.6, Lemma 4].

Proof of Proposition 6.2 : Let k be the dimension of $\underline{G}$, $G := \underline{G}(K)$, the group of K-rational points of $\underline{G}$, and $A := \bigcap_{n \in \mathbb{N}} p^n G$. Then A is a group, its domain is $\wedge$-definable in the separably closed field K and its law is rational over this field. The Zariski closure $\overline{A}$ of A is a variety with a rational map (the addition of A) generically defined on it. We apply the previous claim to it and to the map $f(x_1, ..., x_{p^n}) = \sum_1^{p^n} x_i$, where the $\sum$ refers to the addition in A. Each generic point a of $\overline{A}$ belongs to A, hence, for each integer n, a is of the form $p^n b_n$, for some $b_n \in A$, i.e. $a = f(b_n, ..., b_n)$. By the claim, $a \in K(b_n^{p^n})$, hence $a_{\leq n} \subseteq K(b_n)$. Now $tr(K\langle a \rangle, K) = tr(K(a_\infty), K) = sup\{tr(K(a_{\leq n}), K); n \in \omega\} \leq tr(K(b_n), K) \leq k$. □

References

[Bour] N. Bourbaki, XI, Algèbre, chapitre 5, *Corps commutatifs*, Hermann, Paris 1959.

[Bous] E. Bouscaren, *Proof of the Mordell-Lang conjecture for function fields*, this volume.

[ChKe] C. Chang and J. Keisler, *Model Theory*, North-Holland, Amsterdam 1973.

[CCSSW] Z. Chatzidakis, G. Cherlin, S. Shelah, G. Srour and C. Wood, *Orthogonality of types in separably closed fields*, in Classification Theory (Proceedings of Chicago, 1985), LNM 1292, Springer, 1987.

[CWo] Z. Chatzidakis and C. Wood, manuscript, 1996.

[De 88] F. Delon, *Idéaux et types sur les corps séparablement clos*, Supplément au Bulletin de la SMF, Mémoire 33, Tome 116 (1988).

[Er] Y. Ershov, *Fields with a solvable theory*, Sov. Math. Dokl. 8 (1967), 575-576.

[Hr 90] E. Hrushovski, *Unidimensional theories are superstable*, APAL 50 (1990), 117-138.

[Hr 96] E. Hrushovski, *The Mordell-Lang conjecture for function fields*, J.AMS 9 (1996), 667-690.

[Lan 58] S. Lang, *Introduction to algebraic Geometry*, Interscience Tracts in pure and applied Mathematics, Interscience Publishers, New York 1958.

[Lan1 65] S. Lang, Algebra, Addison-Wesley, 1965.

[Mar] D. Marker, *Zariski geometries*, this volume.

[Me 94] M. Messmer, *Groups and fields interpretable in separably closed fields*, TAMS 344 (1994), 361-377.

[Me 96] M. Messmer, *Some model theory of separably closed fields*, in Model Theory of Fields, Lecture Notes in Logic 5, Springer, 1996.

[Pil] A. Pillay, *Model theory of algebraically closed fields*, this volume.

[We 48] A. Weil, *Courbes algébriques et variétés abéliennes*, Hermann, Paris, 1948.

[Wo 79] C. Wood, *Notes on the stability of separably closed fields*, JSL 44 (1979), 412-416.

[Zie] M. Ziegler, *Introduction to stability theory and Morley rank*, this volume.

Proof of the Mordell-Lang conjecture for function fields

ELISABETH BOUSCAREN

In this chapter we present Hrushovski's model-theoretic proof of the "relative Mordell-Lang conjecture"("The Mordell-Lang Conjecture for function fields" [Hr 96]).

We refer constantly to the previous chapters of this book where all the necessary material was introduced.

As was already explained in the introduction, we have chosen in this volume, for the sake of clarity, to present in full detail the characteristic zero case and to give a self-contained presentation of the model-theoretic tools and results required for this case : Morley rank, ω-stable groups of finite Morley rank. The proof in the characteristic zero case is explained in section 2. The characteristic p case is explained more briefly in section 3. We have also chosen to restrict ourselves to the case of abelian varieties although Hrushovski's proof works for semi-abelian varieties.

Notation: classically, if V is any variety defined over a field k, we denote by $V(k)$ the k-rational points of V. Similarly, suppose that $E \subseteq k^n$ is a definable (or infinitely definable) subset of k^n. Then we denote by $E(k)$ the set of n-tuples in k satisfying E:

$$E(k) = \{\overline{a} \in k^n; k \models E(\overline{a})\}.$$

1 The general statement

Let us recall first the actual statement we are going to prove here, the "relative" or "geometric" case of the Mordell-Lang conjecture. For the background on this and related conjectures, see Hindry's contribution to this volume ([Hi]), for the model theoretic content of the conjecture, see Pillay's second contribution ([Pi2]). For basic definitions and theorems of algebraic geometry, see [Hi] and [Pi1].

Theorem (The Mordell-Lang conjecture for function fields) *Let $k_0 \subset K$ be two distinct algebraically closed fields. Let A be an abelian variety defined over K, let X be an infinite subvariety of A defined over K and let Γ be a subgroup of "finite rank" of $A(K)$. Suppose that $X \cap \Gamma$ is Zariski dense in X and that the stabilizer of X in A is finite. Then there is a subabelian variety B of A and there are S, an abelian variety defined over k_0, X_0 a subvariety of S defined over k_0, and a bijective morphism h from B onto S, such that $X = a_0 + h^{-1}(X_0)$ for some a_0 in A.*

Recall that Γ is said to be of "finite rank" if there exists a finitely generated group Γ_0 such that for every $g \in \Gamma$, there is some $n \geq 1$ (some $n \geq 1$, not divisible by p in the case of characteristic $p > 0$) such that $ng \in \Gamma_0$. We write "finite rank" in quotes in order to avoid confusion with the model-theoretic notion of finite Morley rank or U-rank.

If we do not assume that $X \cap \Gamma$ is dense in X, we get the conclusion that there exist $B_1, \ldots, B_n$ subabelian varieties of A, $S_1, \ldots, S_n$ abelian varieties defined over k_0 and for each i, $1 \leq i \leq n$, Y_i a subvariety of S_i, also defined over k_0 and h_i, a morphism from B_i onto S_i, such that $X = (a_1 + {h_1}^{-1}(Y_1)) \cup \ldots \cup (a_n + {h_n}^{-1}(Y_n))$. If we do not assume that the stabilizer of X is finite, the morphism h has no reason to be bijective. It was explained in [Hi] that the general statement of the conjecture follows from the one above.

We can already note that the assumption that the stabilizer of X is finite (but X is not) implies in particular that h must be non trivial: indeed if S were the trivial abelian variety, then $h^{-1}(X_0) = h^{-1}(S) = B$ and X would be a coset of the (infinite) subabelian variety B of A. But in this case, the stabilizer of X would contain B and be infinite.

In order to give an informal description of the way the proof proceeds, let us go back briefly (and also informally) to the actual meaning of the Lang conjecture.

We have $k_0 \subset K$ two distinct algebraically closed fields, A an abelian variety defined over K, X a subvariety of A defined over K and Γ a subgroup of "finite rank" of $A(K)$.
The aim is to show that either
(1) $X \cap \Gamma$ is a finite union of cosets of subgroups of Γ (finite sets are a union of cosets of the trivial group)
or
(2) we can "descend" the whole situation to k_0 itself, by finding a homomorphism from A to an abelian variety S defined over k_0, such that X is the inverse image (up to translation) of a subvariety of S also defined over k_0.

As we just noted above, what the reduction to the case where the stabilizer of X is finite does for us is to rule out the first case. Note also that with the assumption that $X \cap \Gamma$ is dense in X, then case (1) becomes: X ($X = \overline{X \cap \Gamma}$) is a coset of a subabelian variety of A.

We have seen in [Pi2] that case (1) has an exact (and natural) equivalent statement from the model theoretic point of view: given A and Γ, case (1) will hold for every subvariety X of A if and only if the theory of $(K, +, ., \Gamma, a)_{a \in K}$ is stable and Γ is a one-based group.
So the idea is to show that, given an abelian variety A, either for every "finite rank" subgroup Γ of A, "Γ is one-based" or A "descends" to the smaller field k_0.

The proof will rely heavily on the fundamental dichotomy between one-based groups and non one-based groups which has been one of the focus points of geometric model theory in the past years (see [Las]). But in order to exploit

the rich model theoretic machinery around this dichotomy, we need to replace the group Γ by a **definable** object, while keeping the important condition on Γ that it is in some sense "small" (of finite rank).
We cannot do this by remaining in the context of algebraically closed fields and constructible or algebraic groups: the smallest definable group containing Γ in this context will be the Zariski closure of Γ which will not be "small"(if Γ is the group of torsion elements of A for example, then the closure of Γ is A itself). For this reason, we will **change the class of definable subsets** we are working with.

In the characteristic zero case, following Buium ([Bu 93]), we add new definable sets by adding a derivation on K and then enlarging K to a big algebraically closed field which is also differentially closed. We replace the group Γ by a new definable group H which contains it and has finite Morley rank. We keep trace of the field k_0 by ensuring that it is the field of constants in the differential field K. Note that k_0 then also becomes definable. For the model theory of differentially closed fields, see Wood's contribution ([Wo]).

In the characteristic p case, we do not add new definable sets on the algebraically closed field K itself. We replace K by a (non algebraically closed) separably closed field L, such that all the relevant objects are still present (ie A and X are defined over L, $k_0 = L^{p^\infty} = \cap L^{p^n}$, $\Gamma \subset A(L)$). There are now "more" definable sets because it is no longer true that every definable subset is constructible (the theory of separably closed fields does not eliminate quantifiers in the language of rings). Note that k_0 is now infinitely definable (an infinite intersection of definable subfields) in L. We now consider $H = \cap p^n A(L)$. This will be much "smaller" than $A(L)$: as an infinitely definable set in L, H has finite U-rank. But $X \cap H$ is large enough: $X \cap H$ is also dense in X. For the model theory of separably closed fields, see Delon's contribution ([De]).

From now on the strategy is the same in both cases. The (infinitely) definable group H cannot be one-based: $X \cap H$ would be a finite union of cosets of subgroups of H and hence $X = \overline{X \cap H}$ a coset of some subabelian variety of A, which as we noted above contradicts the assumption that the stabilizer of X is finite. We want to construct the desired homomorphism to get (2). But in order to get some useful information from the assumption of non one-basedness, we need to be in the context of Zariski geometries. The main part of the proof will consist in showing that one can reduce to the case when H is "almost" rank one. Once this reduction is done, we can conclude (Propositions 2.1 and 3.5), using the fact that rank one sets are Zariski geometries in differentially closed fields and in separably closed fields in order to apply the powerful results of Hrushovski and Zilber [HrZi 96], described in Marker's chapter [Mar].

Properties of abelian varietes used in the proof see [Hi]:
Let K be algebraically closed, let A be an abelian variety over K:
(P1) A is a connected commutative algebraic group.
(P2) The torsion of A is infinite, for all n the n-torsion of A is finite.

(P3) A is strongly rigid : if A is defined over $k \subset K$, if G is a closed subgroup of A, then G is defined over the algebraic closure of k, in characteristic 0, and over the separable closure of k in characteristic $p > 0$.
(P4) Suppose that K has characteristic $p > 0$ and that A is defined over $k \subset K$. Let $L \subset K$ denote the separable closure of k; then $A(L)$ is a subgroup of A and $A(L)$ is dense in A. If $a \in A(L)$, if $b = na$, for some n prime-to-p, then $b \in A(L)$.
(P5) (Chevalley's theorem) Let G be an algebraic group defined over K, there is a minimal closed subgroup M of G, defined over K, such that G/M is an abelian variety.

2 The characteristic 0 case

2.1 The model theoretic setting and the dichotomy

In order to replace the group Γ by a definable group, we enrich the structure and pass to a differentially closed field. All facts we will be using about differentially closed fields can be found in [Wo].

Choose δ, a derivation on K, such that k_0 is the field of constants of δ in K. Let L be the differential closure of K (in particular L is algebraically closed), then k_0 is again the field of constants of L and is hence definable in L. Of course, replacing the original K in the statement of the theorem by the bigger field L makes no difference. But we want to work inside a slightly saturated model of the theory of differentially closed fields. So let L' be an $\aleph_0$-saturated elementary extension of L, and let k_0' denote the field of constants of L'; of course $k_0 \subseteq k_0'$. We will prove the theorem with k_0' replacing k_0, and L' replacing K: in L' there are B, a subabelian variety of A, S' an abelian variety defined over k_0', X_0' a subvariety of S' defined over k_0', and a bijective morphism h' from B onto S' such that $X = a_0' + h'^{-1}(X_0')$.

We must now check that we can get back the original statement with L and k_0. We choose here to present a fairly direct model theoretic argument, but it is also easy to argue purely algebraically.

First note that, as L is algebraically closed, by rigidity (P3), the abelian subvariety of A, B is also defined over L.

Now the fact that: " h' is a bijective morphism from B (defined over L), onto an algebraic group S', with a subvariety X_0', both defined over k_0', such that $X = a_0' + h'^{-1}(X_0')$"
is a first order statement in the parameters needed to define h', S, B, X in the differentially closed field L' (k_0' is definable in L'). As $L \preceq L'$, there exist S an algebraic group defined over k_0, X_0 a subvariety of S_0, defined over k_0, and h, a bijective morphism such that $X = a_0 + h^{-1}(X_0)$, for some a_0 in $A(L)$. It follows that S itself must also be an abelian variety (in fact although we did not need to use it here, the fact that S is an abelian variety is also a first order property of the parameters).

So, from now on, without loss of generality, we can assume that $k_0 \subset L$,

where L is an $\aleph_0$-saturated differentially closed field, and k_0 is the constant field of L.
We will say that a set is δ-definable if it is definable in the language of differentially closed fields. We will keep the word definable to mean definable in the pure field language. Recall that, as L is algebraically closed, any definable subset (in the field language) is constructible, i.e. definable without quantifiers.

Now, we replace the group Γ by a "small" δ-definable group, as was done by Buium in [Bu 93]: because Γ is a group of "finite rank", there is a δ-definable subgroup of A, H, containing Γ, which has finite Morley rank (see [Wo, Theorem 4.1]) (A itself will have infinite Morley rank, as a δ-definable set).
Now most of the work in the rest of the proof consists in showing that one can reduce to the case when H is almost strongly minimal (that is, H is contained in the algebraic closure of a strongly minimal δ-definable set) and not one-based (see definition in [Zie, section 5], [Las, section 6]). This will be done in section 2.3, but we will prove immediatly the main proposition, which will enable us to conclude once this reduction is done.

Proposition 2.1 *Let L be an $\aleph_0$-saturated differentially closed field and $k_0 \subset L$, the constant field of L. Let A be any abelian variety defined over L and let H be an almost strongly minimal connected δ-definable subgroup of A. Suppose that H is not one-based.*
(i) Then there exists an abelian variety S, defined over k_0, and a bijective morphism f from $\overline{H}$ (the Zariski closure of H) into S, such that $f(H) = S(k_0)$.
(ii) Let X be a subvariety of A, defined over L, such that $X \cap H$ is dense in X. Then there is subvariety X_0 of S, defined over k_0, such that $X = f^{-1}(X_0)$.

Proof of the Proposition : Let B be a strongly minimal δ-definable set such that $H \subseteq acl(B)$. Easily, if H is not one-based, B cannot be locally modular [Zie, 5.14].

Recall that in differentially closed fields, strongly minimal sets (one may have to take away a finite number of points) are Zariski geometries [Mar, Theorem 6.1]. Now, by the dichotomy theorem for Zariski geometries, as B is not locally modular, B interprets an algebraically closed field [Mar, Theorem 3.3]. This algebraically closed field has finite Morley Rank (it is interpretable in a rank one set) and we know that such a field in L must be δ-definably isomorphic to k_0, the constant field [Wo, Theorem 3.5]. Therefore B and k_0, and hence also H and k_0, are not orthogonal.

It follows ([Zie, Theorem 6.5]), as differentially closed fields have elimination of imaginaries, that there is a δ-definable group G, $G \subseteq {k_0}^k$, definable with parameters from k_0, and a δ-definable homomorphism h with finite kernel which maps H onto G. As H is connected, so is G.

First, a classical argument enables us to reverse the map h. Let $n = |(Ker\ h)|$. Define, for $y \in G$ such that $y = h(x)$, $g_0(y) = nx$. This is well defined, as $h(x) = h(x')$ implies that $x - x' \in (Ker\ h)$, hence $nx = nx'$. Clearly

g_0 is a homomorphism; by finiteness of the n-torsion in H (H is a subgroup of an abelian variety (P2)), g_0 has finite kernel, and g_0 is surjective because H is connected : by finiteness of the n-torsion, H and nH have the same Morley rank (by [Zie, 2.9]), hence nH has finite index in H (see [Las, 2.2]).
Now consider $G_1 = G/(Ker\ g_0)$. The group G is a subset of ${k_0}^m$, δ-definable with parameters from k_0 and $(Ker\ g_0)$ is finite, hence G_1 is again δ-definable (elimination of imaginaries in differentially closed fields) over k_0, $G_1 \subseteq {k_0}^k$ for some k. Consider now the δ-definable isomorphism, g, induced by g_0 from G_1 onto H.

We know that k_0, the field of constants, is a pure field [Wo, 1.10 and 1.11]. So there exists a (connected) algebraic group G_2, defined over k_0, such that $G_2(k_0) = G_1$ and $g_{\restriction G_2(k_0)}$ is rational (over some parameters in L).
The surjective rational group homomorphism g extends naturally to the Zariski closures in L; consider this extension $\overline{g}\ :\ \overline{G_2(k_0)} = G_2(L) \mapsto \overline{H}$; $\overline{g}$ is a surjective morphism. Note that $\overline{H}$ is a closed connected subgroup of A, hence an abelian variety.

We now claim that $(Ker\ \overline{g})$ is trivial : let M be the k_0-definable minimal closed subgroup of G_2 such that G_2/M is an abelian variety (P5). As $G_2/(Ker\ \overline{g})$ is isomorphic to the abelian variety $\overline{H}$, $(Ker\ \overline{g}) \supseteq M$, hence $M(k_0) \subseteq (Ker\ \overline{g})(k_0) = \{0\}$ ($\overline{g}_{\restriction k_0}$ is injective). As M is defined over k_0, M is trivial and G_2 is an abelian variety. But then, by strong rigidity (P3), the closed subgroup of G_2, $(ker\ \overline{g})$, is also defined over k_0, and must also be trivial.

Consider now the inverse map of $\overline{g}$, which we will denote f, from $\overline{H}$ into G_2; f is a bijective morphism from $\overline{H}$, subabelian variety of A, into $S = G_2$, abelian variety defined over k_0 and $f(H) = G_2(k_0)$. This proves (i).

Let $X_0 = f(X)$; then as X is closed irreducible and $\overline{X \cap H} = X$, $f(X)$ is closed, it is irreducible and must be the Zariski closure of $f(X \cap H)$.
Now, $f(X \cap H) \subseteq f(H)$, hence it is contained in k_0. It follows that its Zariski closure is also defined over k_0 itself ([Pi1, 2.13])). Hence $X_0 = f(X) = \overline{f(X \cap H)}$ is defined over k_0. This finishes the proof of the proposition. □

Now in order to give a better idea of the structure of the proof, we will in the next short section, prove a particular case of the Mordell-Lang conjecture, where the extra assumptions enable us to reduce very quickly to the strongly minimal case and to Proposition 2.1.

2.2 A simple case as a warm up

Definition 2.2 *An abelian variety is* **simple** *if it has no proper non trivial subabelian variety (equivalently, if it has no proper non trivial irreducible closed subgroup).*

Theorem 2.3 *Let $k_0 \subset K$ be two algebraically closed fields of characteristic zero. Let A be a simple abelian variety defined over K, let $X \subseteq A$ be a curve defined over K, and let T denote the torsion group of A.*
Then, one of the following holds:

(i) $X = A$

(ii) *$X \cap T$ is finite*

(iii) *there is an isomorphism f from A into S, where S is an abelian variety defined over k_0, and there is X_0, a curve in S, defined over k_0, such that $X = f^{-1}(X_0)$.*

Remark: This implies in particular that if A has trace zero over k_0, and if X has genus at least 2, then $X \cap T$ is finite.
Note also that this follows from the general statement of Mordell-Lang's conjecture and that as X is a curve, if $X \cap T$ is infinite, then it is Zariski dense in X.

Proof: By what we explained in the preceding section, we can suppose that we have a derivation δ on K, that K is an $\aleph_0$-saturated differentially closed field, and that k_0 is the field of δ-constants of K.

By [Wo, Cor. 4.10], as A is simple, there is a connected δ-definable subgroup of A, H, of finite Morley rank, such that $H \supset T$ and H is minimal, that is, H contains no proper infinite δ-definable subgroup.

Claim 2.4 *H is almost strongly minimal and connected.*

Proof of the Claim: Let $B \subseteq H$ be a δ-definable subset, strongly minimal and indecomposable (such a B always exists). Then by Zilber's indecomposability theorem [Las, section 5], for any $b_0 \in B$, the group C, generated by $B - b_0 = \{y - b_0; y \in B\}$, is δ-definable. By minimality of H, $H = C$, and of course $C \subseteq acl(B)$. The connectedness is obvious from the minimality assumption. □

Now there are two cases :
1) B is locally modular. Then the group H is one-based (see [Las, section 6], [Zie, 5.14]).
So we know that every δ-definable subset of H is a boolean combination of cosets of δ-definable subgroups of H. This implies that H is strongly minimal: B itself must be such a boolean combination, but by minimality of H, B must be equal to $H \setminus F$ for some finite F. So, by the strong minimality of H, either $X \cap H$ is finite, hence also $X \cap T$ is finite, this is (ii) or $X \cap H$ is cofinite, hence equal to $H \setminus E$, for some finite E. But then $\overline{H} = A = \overline{(X \cap H) \cup E}$. By irreducibility of A, $A = \overline{X \cap H} = X$. This is (i).
2) B is not locally modular, hence H is not one-based.
We can now apply Proposition 2.1 which, together with the fact that $\overline{H} = A$, by simplicity of A, gives us directly (iii). □

2.3 The general characteristic zero case

As we saw in the first section, we are now reduced to proving the following:

Theorem 2.5 *Let L be an $\aleph_0$-saturated differentially closed field, and let $k_0 \subset L$ be the field of constants of L. Let A be an abelian variety defined over L, let X be a subvariety of A defined over L, with finite stabilizer in A. Let H be a δ-definable subgroup of $A(L)$, with finite Morley rank, such that $X \cap H$ is Zariski dense in X. Then there is a subabelian variety B of A and there are S, an abelian variety defined over k_0, X_0, a subvariety of S defined over k_0, and a bijective morphism h from B onto S, such that $X = a_0 + h^{-1}(X_0)$.*

Note that we can replace X by any A-translate of X. This will be done freely in the course of the proof.

We want to replace the finite Morley rank group H by a sum of almost strongly minimal groups.

By [Las, section 7], there exists a δ-definable subgroup G of H, which has the following properties :

- G is connected, there is some finite F such that $G \subseteq acl(F \cup Y_1 \cup \ldots \cup Y_n)$, with each Y_i strongly minimal δ-definable, and G is maximal such.
- $G = G_1 + \ldots + G_k$, where each G_i is an almost strongly minimal connected δ-definable subgroup of G, and the G_i's are pairwise orthogonal.

Let $F_0 \supseteq F$ denote some finite set in L such that A, X, H, G and all the G_i's are defined over F_0.

Definition 2.6 *Let G be a δ-definable group, defined over F_0. We say that G is* **rigid** *if all δ-definable connected subgroups of G are δ-definable over $acl(F_0)$. Note that G is rigid iff there is no infinite family of uniformly δ-definable subgroups of G.*

Claim 2.7 *The group G is rigid.*

Proof of the Claim : We know that $G = G_1 + \ldots + G_k$, where each G_i is an almost strongly minimal connected subgroup of G, and the G_i's are pairwise orthogonal. We will leave it to the reader to check that it is sufficient to prove that for each i, G_i is rigid.

Now there are two cases:

- G_i is one-based, in this case we know [Las, 6.2] that it is rigid.

- G_i is not one-based. In this case, by Proposition 2.1 (i), there is an abelian variety S, defined over k_0, and a bijective morphism h from $\overline{G_i}$ into S, such that $h(G_i) = S(k_0)$. As $S(k_0)$ is an abelian variety, there is no uniform family of closed subgroups in $S(k_0)$ (P3). Any δ-definable subgroup of $S(k_0)$ is constructible (hence closed [Pil, 4.3]) again because k_0 is a pure algebraically closed field. So $S(k_0)$ is rigid and this is clearly preserved by definable isomorphism. □

Now our assumption is that $X \cap H$ is dense in X. We also know (although this will not be used directly in what follows) that H cannot be one-based: if it were, then $X \cap H$ would be a finite Boolean combination of cosets of subgroups of H, but then $X = \overline{X \cap H}$ would be one coset and, as we noted earlier, this contradicts the fact that the stabilizer of X is finite.

The aim of the next two propositions is to reduce to the almost strongly minimal case in order to apply Prop. 2.1. First (Prop. 2.10), we show that up to translation, $G \cap X$ is dense in X, then (Prop. 2.12), we show that up to translation again, for some i, G_i is not one-based and $G_i \cap X$ is Zariski dense in X.

First let us replace $X \cap H$ by a complete type. Let p be a complete type over L, such that $p \vdash x \in H$. Recall from [Las] that

$$Stab_p = \{h \in H(L); h + p = p\}$$

is a δ-definable subgroup of $H(L)$.

Lemma 2.8 *If q is a complete stationary type over $E \supseteq F_0$, E finite, $q \vdash x \in H(L)$ and if Y denotes the subset $q(L)$, we define* **the (model-theoretic) stabilizer of Y**, *$Stab(Y)$ to be*

$$Stab(Y) = \{h \in H : MR((h+Y) \cap Y) = MR(Y)\}$$

where MR denotes the Morley rank in the sense of the differentially closed field L.
Then if q' denotes the unique non forking extension of q to L, $Stab(Y) = Stab_{q'}$.

Proof : Note first that $h \in Stab_{q'}$ iff for any finite $D \subset L$, for some (= any) b realizing q and independent from Dh over E, then $tp(h+b/EDh) = tp(b/EDh)$, the unique non forking extension of q to EDh. It follows easily that if $h \in Stab_{q'}$, then $MR((h+Y) \cap Y) = MR(Y) = MR(q)$.

Conversely, if D is a finite subset of L, let $a \in (h+Y) \cap Y$ have Morley rank equal to the rank of q over EDh. Then $a = h + c$ for some c realizing q and $MR(tp(a/EDh)) = MR(tp(c/EDh))$, hence by stationarity, $tp(a/EDh) = tp(c/EDh)$. □

Lemma 2.9 *There is a complete stationary type q (over some finite $E \supseteq F_0$), in $X \cap H$, such that $Y = q(L)$ is dense in X, and the (model-theoretic) stabilizer of Y in H is finite.*

Proof : Note that, as X is a closed irreducible set, if $X \cap H = D_1 \cup \ldots \cup D_n$, where D_i is an $acl(F_0)$-δ-definable subset of $X \cap H$, for each i, then as

$$\overline{X \cap H} = X = \overline{D_1} \cup \ldots \cup \overline{D_n},$$

there must be some i such that $X = \overline{D_i}$. Hence there is some complete type $q = \bigcap_{i<\omega} D_i$, over $acl(F_0)$, where we can suppose that the D_i's form a decreasing sequence of δ-definable sets, such that, for each i, D_i is dense in X. It follows that $Y = q(L)$ is Zariski dense in X : we need to check that for any open subset U of X, $Y \cap U \neq \emptyset$. But our assumption is that for each i, $U \cap D_i \neq \emptyset$.

By compactness and saturation (in the theory of differentially closed fields), $U \cap Y \neq \emptyset$. By elimination of imaginaries, q is a stationary type, and by ω-stability, there is some finite $E \supseteq F_0$, $E \subseteq acl(F_0)$, such that $q \restriction E$ is stationary. Now choose E finite, $E \supseteq F_0$, and $q \in S(E)$, $q \vdash x \in (X \cap H)$, such that q is stationary, $q(L) = Y$ is dense in X and q has minimal Morley rank. Note that as X is infinite, $MR(q) = n > 0$.

We claim that the stabilizer of Y in H is finite. Let U be any open subset of X, defined over some finite $D \subset L$. By the minimality assumption on the Rank, $MR(Y \cap U) = n$. Let $h \in Stab(Y)$ and let $a \in Y \cap U$, of Morley rank n over DEh. Then $h + a \in (h + Y) \cap Y$, and $MR(h + a/DEh) = MR(a/DEh)$. By stationarity of q, $tp(a/DEh) = tp(h + a/DEh)$. In particular, $h + a \in U$. This shows that $(h + Y) \cap Y$ is also dense in X. But $\overline{(h + Y) \cap Y} \subseteq (h + X) \cap X$, hence $(h + X) \cap X = X$, which means that $h + X = X$, that is, h belongs to the (setwise) stabilizer of X which is finite by assumption. □

The next proposition is true in general for any commutative finite Morley rank group in a stable theory T. In order to remain within our chosen context in this volume, we suppose that T is ω-stable.

Proposition 2.10 *let T be an ω-stable theory. Let H be a definable commutative finite Morley Rank group, let G be a maximal connected definable subgroup of H such that $G \subset acl(F \cup X_1 \cup \ldots \cup X_n)$, where F is finite and the X_i's are strongly minimal.*
Let $E \supseteq F$ be a finite set such that H, G and the X_i's are defined over E. Suppose that G is rigid , i.e. every definable connected subgroup of G is definable over $acl(E)$.
Let q be a complete stationary type in H, over E such that $Stab(q(L))$ is finite. Then $q(L)$ is contained in a unique coset of G.

Proof : First some notation: if $y \in H$, we will denote by $y + G$ the coset of y considered as a subset of H, and we will denote by $\widehat{y}$ the coset of y, considered as an element of H/G in H^{eq}. Let $Y = q(L)$.

Note that $y+G$ is a subset of H which is definable with parameters in $E \cup \{\widehat{y}\}$. We want to show that Y is contained in one coset of G. By stationarity, q is a complete type over $acl(E)$, hence, for $y \in Y$, all elements $\widehat{y}$ also have the same type over $acl(E)$. So either Y is contained in one coset, or it intersects infinitely many, which exactly means that the Morley rank of $t(\widehat{y}/acl(E))$ is non zero.

Claim 2.11 *Let $y_0 \in Y$; there is some $z \in y_0+G$ such that $z \in acl(E \cup \{\widehat{y_0}\} \cup G)$.*

We first show why the claim finishes the proof.

Let $\widehat{q_0}$ denote the type of $\widehat{y_0}$ over $acl(E)$ and suppose for contradiction that $MR(\widehat{q_0}) > 0$. Choose some finite $F_1 \supseteq E$ and $\widehat{q_1}$, a complete type over F_1 extending $\widehat{q_0}$ of Morley rank one and multiplicity one. As $\widehat{q_1}$ is a complete type over F_1, by the claim, $q_1(\widehat{y}] \vdash$ "$\exists z \in \widehat{y}, \exists v \in G, \theta(z, v, \widehat{y})$", where for any v, w, $\theta(z, v, w)$ is an algebraic formula over $\{v, w\}$ with parameters in F_1.

Now let X_0 be a strongly minimal subset of H/G such that $\widehat{q_1} \vdash \widehat{y} \in X_0$ and $\widehat{y} \in X_0 \vdash$ "$\exists z \in \widehat{y}, \exists v \in G, \theta(z, v, \widehat{y})$".

Hence for any $\widehat{y} \in X_0$, for any $z \in H$, if $z \in \widehat{y}$, then $z + G \subseteq acl(F_1 \cup \{\widehat{y}\} \cup G)$.

Recall [Las, section 5] that any strongly minimal subset contains a strongly minimal indecomposable subset. Hence we can choose X_0 to be indecomposable. Choose some $\widehat{b} \in X_0$. By Zilber's indecomposability theorem [Las, 5.2], the subgroup K of H/G generated by $\{\widehat{x} - \widehat{b}; \widehat{x} \in X_0\}$ is definable and connected. Clearly, if $\widehat{h} \in K$, then $h + G \subseteq acl(F_1 \cup \{\widehat{h}\} \cup G)$.

Now consider the following subgroup C of H, $C = \{h \in H; \widehat{h} \in K\}$. Then $C \subseteq acl(X_0 \cup G \cup F_1)$. Of course $C \supset G$, $C \neq G$, and hence C contradicts the maximality of G.

It remains only to prove the claim.

Proof of Claim 2.11 : Consider U, a non empty subset of $y_0 + G$ definable over $(E \cup \{\widehat{y_0}\} \cup G)$, of minimal Morley rank and degree. We are going to show that U must be finite. The claim follows immediatly.

Note that the minimality of U implies that for any V, nonempty definable (or infinitely definable) set over $(E \cup \{\widehat{y_0}\} \cup G)$, if $V \cap U \neq \emptyset$, then $U \subseteq V$.

Let $S = \{s \in G; s + U = U\}$. We claim that for all $s \in S$, $s + y_0 \in Y$: indeed, as $U \subseteq y_0 + G$, there is some $b \in G$ such that $y_0 \in b + U$. Hence $b + U \cap Y \neq \emptyset$. By the remark above, it follows that $b + U \subset Y$. Now for $s \in S$, $s + y_0 \in s + b + U = b + U \subseteq Y$.

Now let S^0 be the connected component of S [Las]. By rigidity of G, S^0 is definable over $acl(E)$. As Y is a complete type over $acl(E)$, it must also hold that for all $y \in Y$, for all $s \in S^0$, $s + y \in Y$, hence $S^0 \subseteq Stab(Y)$. Hence S^0 is finite and as $[S : S^0]$ is finite, S itself is finite.

But by minimality of U, U must be a coset of S : let $u, v \in U$, then $u - v \in G$; now as $(u - v) + U \cap U \neq \emptyset$, we must have that $(u - v) + U \subseteq U$; it follows that $u - v \in S$.

This means that U is finite hence algebraic over $(E \cup \{\widehat{y_0}\} \cup G)$ and this gives the claim. □

By the previous proposition and lemma 2.9 if we replace X by a translate we can assume that there is a complete stationary type Y over $E \supseteq F_0$ in $X \cap H$ such that $\overline{Y} = X, Y \subseteq G$ and the stabilizer of Y in G is finite.

Recall that $G = G_1 + \ldots + G_k$, where each G_i is almost strongly minimal and the G_i's are pairwise orthogonal. As $Y \subset G$, $\overline{X \cap G} = \overline{Y} = X$. Because the stabilizer of X is finite, again this rules out that G is one-based. Hence one of the G_i's must be non one-based. Our assumption that Stab(Y) is finite (but Y is infinite) implies that Y is contained in a coset of this G_i:

Proposition 2.12 *There is a unique $j_0 \leq k$ such that G_{j_0} is not one-based; Y is contained in a single coset of G_{j_0}.*

Proof : Again, as in the proof of Prop. 2.1, because strongly minimal sets are Zariski geometries in differentially closed fields, if some G_j is not one-based, it must be non orthogonal to the constant field of L, k_0. But, non orthogonality is an equivalence relation for strongly minimal sets [Zie, 6.4], hence two such non one-based almost strongly minimal groups cannot be orthogonal, so there is at most one non one-based G_j.
We know that $Y \subseteq G_1 + \ldots + G_k$. Write G as $B_1 + G_{j_0}$, where B_1 is the sum of all the one-based G_i's, and G_{j_0} is not one-based.
By orthogonality, we must have that $Y = U + V$, where U is a complete type over $acl(E)$ in B_1, and V is a complete type over $acl(E)$ in G_{j_0}. Now, B_1 is a one-based group (B_1 is a sum of one-based groups), and it follows from the fact that the stabilizer of Y is finite, that the stabilizer of U in B_1 must also be finite. But [Las, 6.2], in a one-based group, a complete type is a coset of its stabilizer, hence U must be finite, that is, $U \subseteq acl(E)$; as Y is a complete type over $acl(E)$, U must be reduced to one element. Hence $Y \subseteq u + G_{j_0}$ for some $u \in G$. □

Let G_0 denote the unique non one-based almost strongly minimal group given by the above proposition. By translating again we can suppose that $Y \subseteq G_0$, and hence $\overline{Y} = \overline{X \cap G_0} = X$.
We now have the right assumptions to apply Proposition 2.1: there exists an abelian variety S, defined over k_0, X_0, a subvariety of S also defined over k_0, a bijective morphism f, from $\overline{G_0}$ into S, such that $X = f^{-1}(X_0)$.
This finishes the proof of Theorem 2.5. □

3 The characteristic p case

We will now explain the model theory setting for the characteristic p case without going into all the details. The strategy will be similar to the one used in the characteristic zero case, except that here one works with the theory of separably closed fields, which is not ω-stable, but only stable. Hence there is no notion of rank in this case which applies to all definable sets, like the Morley rank in the case of an ω-stable theory. But certain infinitely definable subsets (that is, subsets defined by an infinite conjunction of formulas) will carry a natural notion of dimension or rank and the properties we have proved in previous chapters about groups of finite rank, one-basedness... can be proved in exactly the same way, although in this volume we have chosen, for the sake of clarity to present only the case of ω-stable theories. The rank replacing the Morley rank is the U-rank or Lascar rank. The sets playing the role of the strongly minimal sets are the complete minimal types: let M be a κ-saturated model, and Q an infinitely definable subset of M^n (defined over some set of parameters of cardinality strictly smaller than κ); we say that Q is minimal if for any definable set $D \subseteq M^n$, $Q \cap D$ is finite or cofinite in Q. These complete types are exactly the complete types of U-rank one. It is also true that minimal complete types in separably closed fields of finite degree of imperfection are Zariski structures, (with an adequate slightly adapted definition, see [De], [Mar, Section 5]).

The main differences with the characteristic zero case will be of three kinds:
- we will be working inside a separably closed field L which is not algebraically closed.
- We will be considering infinitely definable subsets instead of definable subsets of this separably closed field. In particular the group Γ will be replaced by an infinitely definable subgroup H of $A(L)$, but there is one advantage over the characteristic zero case, as we will see, one does not need to appeal to a construction as complicated as the Manin-Buim one, and H is easily described.
- Finally, but this turns out not to be a serious problem (see the proof of Proposition 3.5), in characteristic p, we have many definable maps which are not locally rational, but p-rational or p-regular (that is, the composition of a rational map with the negative power of the froebenius (see [Pi1, 4.8])).

We will explain fully the model theory setting and point out the main steps along in the proof where this different setting requires a little more work.
All the facts and definitions needed from the theory of separably closed fields can be found in Delon's contribution to this volume [De].

The model theoretic setting and the dichotomy:

We now go back to the statement of the Mordell-Lang conjecture for function fields (Section 1) and suppose that K is an algebraically closed field of characteristic $p > 0$. As Γ has "finite rank", there is a finitely generated group Γ_0 such that for all $g \in \Gamma$, there is some $n \geq 1$, such that p does not divide n and such that $ng \in \Gamma_0$.
Now let K_0 be a subfield of K, finitely generated over k_0, $k_0 \subset K_0 \subset K$, such that A and X are defined over K_0, and the generators of Γ_0 are K_0-rational points of A.
Note that K_0 has finite degree of imperfection and that $k_0 = K_0^{p^\infty}$. Let L denote the separable closure of K_0. Then again $k_0 = L^{p^\infty}$.
Note that $A(L)$ is a subgroup of $A(K)$, and that Γ is also a subgroup of $A(L)$ (P5).
It is clear that we can replace the original algebraically closed field K by $\tilde{L}$, the algebraic closure of L, in the statement of the theorem.
We are now going to work in the theory of separably closed fields of fixed finite degree of imperfection.
As with differentially closed fields, we go to some saturated extension: let L' be an $\aleph_1$-saturated elementary extension of L, and let k_0' denote the field L'^{p^∞} (because we are working with a stable non ω-stable theory and we are considering infinitely definable objects, we need to take an $\aleph_1$-saturated extension).
Here one has to be a little more careful : L' itself is not algebraically closed, so we work inside the algebraic closure of L', and when we talk about abelian varieties, or Zariski closures etc., we mean them in the algebraic closure.
As in characteristic zero, we are going to work in L' and show that there are B,

a subabelian variety of A, S' an abelian variety defined over k'_0, X'_0 a subvariety of S' defined over k'_0, and a bijective morphism h' from B onto S' such that $X = a'_0 + h'^{-1}(X'_0)$. And we need to see that we can here also get back the original statement with k_0 instead of k'_0.
By rigidity of abelian varieties (P3), as L is separably closed, B is in fact defined over L.
Let $\widetilde{L}'$ denote the algebraic closure of L'. Then the statement above is really an elementary statement in the language of fields, in the pair of algebraically closed fields, $(k'_0, \widetilde{L}')$. As $L \preceq L'$, k'_0 and L are linearly independent over k_0 [De].
This implies that the pair of algebraically closed fields $(k'_0, \widetilde{L}')$ is an elementary extension of the pair $(k_0, \widetilde{L})$ (in the language with a new predicate added for the small field).
The fact that: "there is a bijective morphism h', from B (defined over $\widetilde{L}$) onto an algebraic group S', with a subvariety X'_0, both defined over k'_0, such that $X = a'_0 + h'^{-1}(X'_0)$"
is a first order statement in the parameters needed to define h', B, X, S, X_0, in the pair of algebraically closed fields $(k'_0, \widetilde{L}')$. As $(k_0, \widetilde{L}) \preceq (k'_0, \widetilde{L}')$, it holds also in $(k_0, \widetilde{L})$.

From now on we can suppose that L is an $\aleph_1$-saturated separably closed field. We work in the theory of separably closed fields of fixed finite degree of imperfection, in the language described in [De], in which we have elimination of imaginaries and elimination of quantifiers. When we talk of "definable subsets of L", we mean in the sense of the theory of separably closed fields. We will use the word constructible to mean a subset definable without quantifiers in the pure field language. Recall that in any algebraically closed field, so in particular in $\widetilde{L}$, the algebraic closure of L, any definable subset is constructible.

We are not here going to replace the group Γ by a definable group which contains it but by a "small" (of finite rank in the theory of separably closed fields) infinitely definable subgroup H of $A(L)$ which will still be such that that $H \cap X$ is Zariski dense in X (the Zariski closures are taken in the algebraic closure of L). A natural candidate is $H = \bigcap_{n \in \omega} p^n A(L)$; H does indeed have finite rank [De, Prop.6.2] and the fact that Γ has "finite rank" and $\overline{\Gamma \cap X} = X$ ensures that (up to translation) $\overline{H \cap X} = X$ (this is 3.2 below).

Again our assumption that X (and hence $H \cap X$) is infinite but has finite stabilizer rules out that the group H is one-based. From this we want to conclude that there is a bijective morphism from a subabelian variety of A into an abelian variety defined over $k_0 = L^{p^\infty}$. As in the case of characteristic zero and differentially closed fields, we are going to use the fact that rank one sets in separably closed fields are Zariski geometries and that the only infinitely definable finite rank field in L is k_0.

First let us check that indeed $\overline{H \cap X} = X$.

We are first going to need the following useful explanation of what it means for an infinitely definable subset of L to be Zariski dense in a closed set:

Claim 3.1 *Let L be an $\aleph_1$-saturated separably closed field , $\tilde{L}$ the algebraic closure of L, and X any irreducible (Zariski) closed set defined over L. Let F be any finite subset of L over which X is defined.*
1) If Y is an infinitely definable subset of L, over F, then Y is Zariski dense in X iff, for all $n < \omega$, there are $a_1, a_2, \ldots, a_n$ in Y, such that, in $\widetilde{L}$, $(a_1, a_2, \ldots, a_n)$ are n independent realizations of the generic of X, in the sense of algebraically closed fields.
2) If $Y = \bigcap_{i<\omega} Y_i$, where for each i, Y_i is a definable subset of L over F and Y_i is Zariski dense in X, and if the Y_i's form a decreasing sequence, then Y is Zariski dense in X.

Proof of the Claim : First we note that 2) follows from 1) by compactness in the theory of separably closed fields (by irreducibility of X, the type of n independent generic points of X, in the sense of algebraically closed fields is uniquely determined).
For 1), suppose that Y is dense in X. This implies in particular that for any open subset U of $\widetilde{L}$, defined over some finite $F' \subset L$, $F \subseteq F'$, $Y \cap U(L)$ is non empty. By compactness and saturation (in L) this means that

$$Y \cap (\bigcap \{U(L); U \text{ open non empty } F'\text{-definable subset of } \widetilde{L}\}) \neq \emptyset.$$

Now we show by induction on n that there are $a_1, \ldots, a_n$ in Y realizing n independent generics of X over F.
For $n = 1$, the generic type q of X (over F) is the intersection of all F-definable open subsets of X [Pi1, 2.8]. Hence there must be some a_1 realizing q in Y. Now suppose we have $a_1 \ldots, a_n$ in Y; a_{n+1} will realize the generic of X over $F \cup \{a_1, \ldots, a_n\}$ if it is contained in the intersection of all $F \cup \{a_1, \ldots, a_n\}$-definable open subsets of X in $\widetilde{L}$. We have seen that this intersection must meet Y.
Conversely, let U be any open subset of X in $\widetilde{L}$, we want to show that $Y \cap U \neq \emptyset$. Let $F' \supseteq F$ be finite such that U is defined over F', $F' \subset \widetilde{L}$. By assumption, for any n we can find $a_1, \ldots, a_n$ in Y, independent realizations of the generic type of X over F. By ω-stability of $\widetilde{L}$, as F' is finite, if n is big enough, there must be some a_i such that a_i and F' are independent over F [Zie]. Hence a_i realizes the generic type of X over F' and must be contained in U. □

Claim 3.2 *Let L be an $\aleph_1$-saturated separably closed field of finite degree of imperfection, A an abelian variety defined over L, X a subvariety of A defined over L and Γ a "finite rank" subgroup of $A(L)$ such that $\Gamma \cap X$ is Zariski dense in X. Let $H = \bigcap_{n\in\omega} p^n A(L)$. Then there is some coset of H, C, such that $X \cap C$ is Zariski dense in X.*

Proof of the Claim : Let H_n denote $p^n A(L)$, i.e. $H = \bigcap_{n \in \omega} H_n$. As Γ has "finite rank", it follows that for all n, $\Gamma / p^n \Gamma$ is finite, hence Γ is contained in a finite union of cosets of H_n. It follows, as $X \cap \Gamma$ is dense in X and X is irreducible, that there is some coset C_n of H_n such that $X \cap C_n$ is dense in X.
By compactness and saturation, using the previous claim, we can find some $a \in X$ such that, for each n, $C_n = a + H_n$ and hence such that $(a + H) \cap X$ is dense in X. □

Now, by translating X, we can suppose that $H \cap X$ is dense in X and it suffices to prove:

Theorem 3.3 *Let L be an $\aleph_1$-saturated separably closed field of finite degree of imperfection, and let $k_0 = L^{p^\infty}$. Let A be an abelian variety defined over L, let X be an infinite subvariety of A defined over L, with finite stabilizer in A. Let H be an infinitely definable subgroup of $A(L)$, with finite rank in L, such that $X \cap H$ is Zariski dense in X. Then there is a subabelian variety B of A, S, an abelian variety defined over k_0, X_0 a subvariety of S defined over k_0 and a bijective morphism h from B onto S, such that $X = a_0 + h^{-1}(X_0)$.*

Again, most of the work in the proof, exactly as in the characteristic zero case, consists in showing that one can reduce to the case when H is contained in the algebraic closure of a minimal type (we say that H is semi-minimal).
We will not give the details of this part of the proof. One needs to prove the "finite U-rank" versions of the properties of finite Morley rank groups which are explained in Lascar's chapter ([Las]).

It should be noted that, in the case of an abelian variety and of the particular group H we considered above, ie $H = \bigcap_{n \in \omega} p^n A(L)$, by proceeding a little differently, one can in fact avoid the rather heavy Proposition 2.10, or more accurately its equivalent for groups of finite U-rank. This is explained in [Pi 97]. This same simplification does not seem to be possible in the case of a semi-abelian variety.

We are now going to give the full proof of the adequate version of Proposition 2.1 for separably closed fields, which enables us to conclude, once the reduction to a semi-minimal group is done. This is Proposition 3.5.
The main facts about separably closed fields used in the proof of Prop. 3.5 are the following (see [De]):
(S1) $k_0 = L^{p^\infty}$ is infinitely definable, and is a pure algebraically closed field: let F be a definable subset of L^n, with parameters from L,

1. by stability, $F \cap {k_0}^n$ is defined with parameters from k_0.
2. There is a constructible (ie definable without quantifiers in the field language) subset of ${k_0}^n$, E such that $E = F \cap {k_0}^n$. We will allow ourselves to say inaccurately that $F \cap {k_0}^n$ is a constructible subset of ${k_0}^n$.

3. If h is a definable map from $E = F \cap {k_0}^n$ in L^k, defined over $A \subset L$, there is a partition of $E = C_1 \cup \ldots \cup C_m$ where each C_i is a constructible subset of ${k_0}^n$, such that $h_{\restriction C_i}$ is p-rational (over A). We will say that $h_{\restriction E}$ is locally p-rational.

(S2) Any infinitely definable field of rank one in L is definably isomorphic to L^{p^∞}.

(S3) A minimal type is a Zariski type in the sense of [Mar], Section 5.

Definition 3.4 *Let M be an $\aleph_1$-saturated model. Let E be an infinitely definable subset of M^k, defined over some finite subset D. We say that E is* **semi-minimal** *if for some finite $F \supseteq D$, there is a complete minimal type $q \in S(F)$ such that $E \subset acl(F \cup q(M))$.*

Proposition 3.5 *Let L be an $\aleph_1$-saturated separably closed field of finite degree of imperfection, $k_0 \subset L$, $k_0 = L^{p^\infty}$. Let $\bar{L}$ denote the algebraic closure of L. Let A be an abelian variety defined over L. Let H be a semi-minimal infinitely definable connected subgroup of A. Suppose that H is not one-based.*
(i) Then there exists an abelian variety S, defined over k_0, and a bijective morphism f from $\overline{H}$ (the Zariski closure of H in $\bar{L}$) into $S(\bar{L})$, such that $f(H) = S(k_0)$.
(ii) Let X be a subvariety of A defined over L such that $X \cap H$ is dense in X. Then there is subvariety X_0 of S, defined over k_0, such that $X = f^{-1}(X_0)$.

Proof of the Proposition : Let q be a minimal complete type such that $H \subseteq acl(F \cup q(L))$. Easily, if H is not one-based, the minimal type q cannot be locally modular.
Recall that in separably closed fields of finite degree of imperfection, minimal types are Zariski types (S3). By the dichotomy theorem for Zariski types ([Mar, Theorem 5.1]), as q is not locally modular, q interprets an infinitely definable algebraically closed field of rank one. Such a field must be definably isomorphic to k_0 (S2). Hence q and k_0 are not orthogonal.
As in the characteristic 0 case, it follows that there is a connected definable group $G \subseteq k_0$, and a definable surjective homomorphism with finite kernel from H onto G.
But the proof is slightly more complicated because both H and k_0 are infinitely definable and not definable.

Claim 3.6 *There is a connected algebraic group G_2 defined over k_0, and a definable map g, such that $g_{\restriction G_2(k_0)}$ is locally p-rational (over some parameters from L) and $g_{\restriction G_2(k_0)}$ is an isomorphism onto H.*

Proof of the Claim : First we show a version of Theorem 6.5 in [Zie], adapted to our case.

As q and k_0 are not orthogonal, q is a minimal type and $H \subseteq acl(F \cup q)$, over some finite set b (where b contains F and the parameters needed to define

q), $H \subseteq acl(b \cup k_0)$. In particular, consider p the generic type of the connected group H. Without loss of generality the type p is a stationary type over b, and the abelian variety A is also defined over b. There is a formula $\theta(x, y, b)$ such that:
- for each $h \in H$, realizing p, there is some $a \in {k_0}^n$ such that $L \models \theta(h, a, b)$
- for every y, if $\theta(x, y, b)$ holds, then x is algebraic over yb.
Now for $h \in H$, generic over b, consider the following subset of ${k_0}^n$:

$$C(h, b) = \{a \in {k_0}^n; \theta(h, a, b)\}.$$

$C(h, b)$ is a definable subset of ${k_0}^n$, by (S1) above, it is definable with parameters from k_0, and is constructible (that is definable without quantifiers in the pure field language in k_0). Algebraically closed fields have elimination of imaginaries [Pil], so there is some $\tilde{h} \in {k_0}^k$ which is a canonical parameter for the set $C(h, b)$ in k_0. This means that for any automorphism f of k_0, f fixes $\tilde{h}$ iff f fixes $C(h, b)$ setwise. But certainly if f' is any automorphism of the separably closed field L, then the restriction of f' to k_0 is an automorphism of k_0, hence any automorphism of L fixes $\tilde{h}$ iff it fixes $C(h, b)$ setwise. It follows that in L

$$\tilde{h} \in dcl(h, b) \text{ and } h \in acl(\tilde{h}, b).$$

Hence there is in L a definable finite to one map l, with parameters b, such that $l_{\restriction p(L)}$ is onto $\tilde{p}(k_0) \subseteq {k_0}^k$, where $\tilde{p} = tp(\tilde{h}/b)$ is a complete type. Note that there is no reason why, for an arbitrary non generic $h \in H$, we should have that $l(h) \in {k_0}^k$.

Now we proceed exactly as in the proof of Theorem 6.5 in [Zie].
Consider

$$K = \{a \in H; l(a + h) = l(h) \text{ for one } (= \text{all}) \ h \in H \text{ generic over } a\}.$$

Recall that if h is generic over a, $a + h$ is also generic and note that K is a finite subgroup of H.
Now as H is semi-minimal, it has finite rank (U-rank), say r, and it follows exactly as in the case of an ω-stable theory that if one chooses $h_0, \ldots, h_{2r}$ independent realisations of the generic in H, then for any $a_1, a_2 \in H$, there is at least one j such that h_j remains generic over $\{a_1, a_2\}$.
Define $l' : H \to (L^k)^{2r+1}$ by

$$l'(a) = (l(a + h_0), \ldots, l(a + h_{2r})).$$

Then, l' is definable with parameters $\{b, h_0, \ldots, h_{2r}\}$, and $l'(a_1) = l'(a_2)$ implies that $a_1 - a_2 \in K$.
Define $f(a) = \{(l'(a + d); d \in K\}$. By elimination of imaginaries, $f(a) \in L^m$ for some m and by defnition $f(a) = f(a')$ iff $a - a' \in K$. If we let $G_0 = f(H)$, then one can definably equip G_0 with a group structure isomorphic to H/K; we will also denote the group operation on G_0 by $+$. The connected infinitely definable group G_0 is not a priori contained in ${k_0}^m$, but its generic type (the image of the generic type of H/K) is contained in ${k_0}^m$. It is well-known that a

stable connected group can be recovered from pairs of generics. We are going to reprove it here in our specific context.

Let p denote the generic of G_0 . As we noted above, $p \vdash x \in {k_0}^{k(2r+1)}$ and f maps the generic type of H/K onto p.
By stability, in H/K every element is the sum of two generics, more precisely, let $a \in H/K$, let c be any generic of H/K, independent from a, then $a - c = d$ is also generic. Consider now the following definable equivalence relation on $p \times p$:

$$(x, y)R(x', y') \text{ iff } x + y = x' + y'.$$

By (S1), both p and R are defined with parameters from k_0. By elimination of imaginaries in k_0, let $G_1 \subset {k_0}^r$ be definably isomorphic to $p \times p/R$.
Then the following is a definable bijection from G_0 onto G_1: for $a \in G_0$, let $s(a) = d \in G_1$, where $d = (c, a - c)/R$ for some (any) $c \in G_0$, generic independent from a (the definability follows as usual from the definability of types in a stable theory). Through s, transport the group structure on G_1 and let h denote the corresponding definable group homorphism (with kernel K), from H onto G_1. Now by (S1,S2) and [Las, 4.7] (in an ω-stable theory, every infinitely definable group is definable), there is a constructible group G defined over k_0 such that $G \cap k_0 = G_1$.

As in the proof of Prop. 2.1, we can reverse the map h: there is a definable homorphism g_0 from $G \cap k_0$ onto H, with finite kernel.
Consider $G' = (G \cap k_0/(Ker\ g_0))$. As $(Ker\ g_0)$ is finite and contained in k_0, by elimination of imaginaries again and (S1) again, G' is contained in k_0 , is definable over k_0 and is the trace on k_0 of a constructible group. Consider now the definable isomorphism, g, induced by g_0 from G' onto H. Then (S2) $g_{\restriction G'}$ is locally p-rational over some parameters from L.
By the model theoretic version of Weil's theorem [Pi1], there is a (connected) algebraic group G_2, defined over k_0, such that $G_2(k_0)$ (the k_0-rational points of G_2) is constructibly isomorphic to G'.
This gives the statement of the claim. □

Now consider the Zariski closures of $G_2(k_0)$ and H in $\tilde{L}$, the algebraic closure of L. Note that, as k_0 is algebraically closed, $G_2(\tilde{L})$ is the Zariski closure of $G_2(k_0)$, and let $\overline{H}$ denote the Zariski closure of H.
The map $g_{\restriction G_2(k_0)}$ is not a priori a morphism in the algebraic sense, but it is certainly continuous (for the Zariski topology). Now consider, $\overline{g}$, its natural extension to the Zariski closures. Note that $\overline{H}$ is a closed connected subgroup of A, hence an abelian variety, and that $\overline{g}$ is still a surjective p-rational group homomorphism.
Exactly as in the proof of Prop. 2.1, in the characteristic zero case, $(Ker\ \overline{g})$ is trivial (the proof of this fact involves only the pair of algebraically closed fields $k_0 \subset \tilde{L}$).
So in fact, G_2 is an abelian variety, and $\overline{g}$ is bijective. Now consider the inverse map of $\overline{g}$, f, a bijective p-rational homomorphism from $\overline{H}$ onto G_2, abelian variety defined over k_0. Compose f with some adequate power of the Frobenius

map, $f' = x^{p^n} \circ f$; let S denote the image of G_2 under this map f'. Then f' is now a rational bijective morphism from $\overline{H}$ onto S. S is still defined over k_0, and as k_0 is algebraically closed, we still have that $f'(H) = S(k_0)$. This proves (i). The proof of (ii) goes through exactly as in the characteristic zero case. □

References

[Bu 93] A. Buium, *Effective bounds for geometric Lang conjecture*, Duke J. Math. 71 (1993), 475-499.

[De] F. Delon, *Separably closed fields*, this volume.

[Hi] M. Hindry, *Introduction to abelian varieties and the Lang Conjecture*, this volume.

[Hr 96] E. Hrushovski, *The Mordell-Lang conjecture for function fields*, Journal AMS 9 (1996), 667-690.

[Hr] E. Hrushovski, *Proof of Manin's theorem by reduction to positive characteristic*, this volume.

[HrZi 96] E. Hrushovski and B. Zilber, *Zariski Geometries*, Journal AMS 9 (1996), 1-56.

[Las] D. Lascar, *ω-stable groups*, this volume.

[Mar] D. Marker, *Zariski geometries*, this volume.

[Pi 97] A. Pillay, *Model theory and diophantine geometry*, Bull. Am. Math. Soc. 34 (1997), 405-422.

[Pi1] A. Pillay, *Model theory of algebraically closed fields*, this volume.

[Pi2] A. Pillay, *The model-theoretic content of Lang's Conjecture*, this volume.

[Wo] C. Wood, *Differentially closed fields*, this volume.

[Zie] M. Ziegler, *Introduction to Stability theory and Morley rank*, this volume.

Proof of Manin's theorem by reduction to positive characteristic

Ehud Hrushovski

We explain in this note how to deduce a characteristic 0 Mordell-Lang statement for function fields from the positive characteristic version. See the contributions of Bouscaren and Hindry to this volume for the general statement of Mordell-Lang. (See also [Lan 91] for the history and further references.) While we see no obstacle to proving the general statement by the same method, we will restrict the statement to abelian varieties and rational points.

Let K be a finitely generated field of characteristic 0. We will say that an abelian variety A over a field K ***has no isotrivial factors*** if there are no nonzero homomorphisms (defined over $\widetilde{K}$) from an abelian variety defined over $\widetilde{\mathbb{Q}}$ to A. Here $\tilde{k}$ denotes the algebraic closure of a field k.

Theorem Let A be an abelian variety over K, with no isotrivial factors. Let X be a K-subvariety of A. Then

$$X(K) = C(K)$$

where C is a finite union of translates of connected group subvarieties of A.

When X is a curve of genus ≥ 2, by considering an embedding of X in its Jacobian, one can deduce Manin's theorem [Man]: if X does not descend to $\widetilde{\mathbb{Q}}$, then $X(K)$ is finite. The possibility of such a proof is raised in [Lan 91], following the description of Voloch's theorem.

The need to find ordinary reductions of an abelian variety disappears when one knows the general case of Mordell-Lang in positive characteristic. An issue that remains is to ensure that the reduction mod p of a non-isotrivial abelian variety remains isotrivial.

Another issue arises upon generalization to higher dimensions: one must show that a Zariski dense set of points of an abelian variety, specializes to a Zariski dense set of points in the reduced variety. Neron has shown how to ensure that an infinite set of points remains infinite upon reduction, and we skirt the general problem, by using Neron's lemma for various quotient abelian varieties, and using the statement of Mordell-Lang itself.

Neron's idea (in connection with reductions of abelian varieties from function fields to number fields, cf. [Lan 83]) relies on a simple lemma concerning free abelian groups. Here is a somewhat more general statement.

Author partially supported by a grant from the NSF.

Lemma 1 *Let $r : B \to C$ be a homomorphism of abelian groups, $l > 0$. Assume $\cap_n l^n B = (0)$, C has no l-torsion, and r induces an injective map $(B/lB) \to (C/lC)$. Then r is injective.*

Proof: Let $b \in B$, $b \neq 0$, and let $k > 0$ be smallest such that $b \notin l^k B$; then $b = l^{k-1}a$ for some $a \in B \setminus lB$. By assumption, $r(a) \neq 0$, which implies that $l^{k-1}r(a) = r(b) \neq 0$. □

We begin by stating Neron's elementary lemma. We then briefly describe the process of reduction mod p in general, followed by a description of tricks to ensure non-isotriviality. We then note a certain uniformity in the definition of subgroups of abelian varieties that answers the Zariski density problem above. Finally we give a proof of the theorem. The methods are all standard. Most of the lemmas are known, and are presented for expository purposes (with a slight logic slant.)

Reduction of function fields to positive characteristic Let K be a finitely generated field of characteristic 0; it has the form $K = L(V) = L(c)$, with L a number field, V an (absolutely) irreducible smooth affine variety over L, and c a generic point of V over L (in the sense of Weil's Foundations).

Let $\mathcal{O}_L$ be the ring of integers of L and lift the definition of V to $\mathcal{O}_L$. (This can be done in more than one way, but any two ways yield isomorphic reduced varieties modulo all but finitely many primes.) Let p be a prime of L, $\overline{L} = \mathcal{O}_L/p$, $\overline{V} = \overline{V}_p = V \otimes_{\mathcal{O}_L} \overline{L}$ the reduced variety.

We will freely use the "first-order Lefschetz principle". This states that for any first-order sentence σ, σ holds in an algebraically closed field of characteristic 0 iff it holds in algebraically closed fields of arbitrarily large positive characteristic iff it holds in all but finitely many characteristics. Similarly, if σ is a sentence involving parameters from $\mathcal{O}_L$, and σ holds in $\tilde{L}$, then for almost all primes p, σ holds (with the "reduced" parameters) in $\tilde{\overline{L}}$. (And if the parameters are from a regular extension L' of $\mathcal{O}_L$, then for almost all p σ holds in any algebraically closed field containing $\overline{L} \otimes_{\mathcal{O}_L} L'$.) This follows from the fact (Tarski-Chevalley) that every formula in the language of field is equivalent, uniformly in all algebraically closed fields of large enough characteristic, to a quantifier-free formula.

The following is classical (E. Noether, [Noe]); a model theoretic proof was given in [vdDS]; see other references there.

Lemma 2 *For all but finitely many primes p, $\overline{V}$ is irreducible, and* $\dim(V) = \dim(\overline{V})$.

We assume from now on that $\overline{V}$ is irreducible. Let $\overline{K} = \overline{L}(\overline{V})$. Extend the place $L \to \overline{L}$ to a place $K \to \overline{K}$; let R be the corresponding valuation ring.

Let U be a non-principal ultrafilter on the primes of L. Taking the ultraproduct of the maps $r_p : R \to \overline{K}_p$, we obtain a map $r_{p^*} : R \to \overline{K}_{p^*}$, where $\overline{K}_{p^*}$

denotes $Ult_U\overline{K}_p$. Any element of the kernel of r_{p^*} is contained in the kernel p of r_p for almost all p, hence is zero. Thus r_{p^*} injects R into $Ult_U\overline{K}_p$, and extends to an embedding of K in Ult_UK_p.

Lemma 3 *$\tilde{K}$ is linearly disjoint from $\widetilde{\mathbb{Q}}\overline{K}_{p^*}$ over $K\widetilde{\mathbb{Q}}$.*

Proof: First assume K' is a finite extension of K, and K' is linearly disjoint from $\widetilde{\mathbb{Q}}$ over L. In this case we show that K' is linearly disjoint from $Ult_U\overline{K}_p$ over K. Indeed we may write $K = L(V)$, $K' = L(W)$, where V,W are absolutely irreducible varieties over L, and there exists a rational map $f : W \to V$, surjective and of degree $l = [K' : K]$, defined over L. By Lemma 2, for almost all primes p of L, $\overline{W}$ and $\overline{V}$ are irreducible. By the Lefschetz principle, for almost all p, the reduction of f mod p has degree l. Thus $W \times_K Ult_U\overline{K}_p$ and $V \times_K Ult_U\overline{K}_p$ are irreducible, and the map between them has degree l. So

$$[(Ult_U(\overline{K}_p)(W) : (Ult_U(\overline{K}_p)(V)] = l = [K' : K]$$

proving the linear disjointness.

In general, we may take a finite extension K' of K, linearly disjoint from $\widetilde{\mathbb{Q}}$ over $L' = K' \cap \widetilde{\mathbb{Q}}$, and show K' is linearly disjoint from $L'\overline{K}_{p^*}$ over KL'. We can lift almost the primes p of L to primes p' of L', to obtain $Ult_U\overline{KL'}_{p'} = L'Ult_U\overline{K}_p$. Then we reduce to the previous case, applied to $L', L'K, K'$ in place of L, K, K'. □

Let A be an abelian variety over $K = L(V)$. We can reduce as above and obtain a variety $\overline{A}$ over $\overline{K} = \overline{L}(\overline{V})$. For all but finitely many p, $\overline{A}$ is an abelian variety over $\overline{K}$. Indeed we can take A to be projective; then $\overline{A}$ is projective. Moreover by the first-order Lefschetz principle, the group law on A reduces to a group law on $\overline{A}$, for almost all p.

The ring homomorphism $R \to \overline{K}$ induces a group homomorphism $r : A(K) = A(R) \to A(\overline{K})$. We will refer to r as the reduction homomorphism.

Finally, we make the assumption of no isotrivial factors. By the Lang-Neron theorem ([Lan 83], p. 139, Theorem 2), $A(\widetilde{\mathbb{Q}}K)$ is finitely generated.

The following lemma is the mixed-characteristic version of what is done in [Lan 83], chapter on Hilbert Irreducibility, "applications to abelian groups". The Hilbert Irreducibility theorem can be replaced here by the elementary Lemma 2; thus we get "almost all primes" rather than "infinitely many primes".

Lemma 4 *Let A be an abelian algebraic group over K, with $A(\widetilde{\mathbb{Q}}K)$ finitely generated. Then for almost all p, r is injective on $A(K)$*

Proof:

1) Since $A(\widetilde{\mathbb{Q}}K)$ is finitely generated, we can replace K by a finite extension KL', $L' \subset \widetilde{\mathbb{Q}}$, so that $A(K) = A(\widetilde{\mathbb{Q}}K)$. Note that $A(K)$ has a finite torsion subgroup.

2) For almost all p, r is injective on the torsion points of $A(K)$.

This is immediate, since we are dealing with a finite group. (Any two distinct points have two distinct coordinates in some affine embedding of an open subset;

for all but finitely many primes, the embedding does not involve division by zero, and the two coordinates remain distinct.)

3) Fix a prime l such that $A(\tilde{\mathbb{Q}}K)$ has no points of order l. For large enough p, $A(\overline{K})$ has no points of order l.

Suppose otherwise: then $A(\overline{K}) = A(\overline{K}_p)$ has an order l-point c_p , for infinitely many primes p of L. Let U be a non-principal ultrafilter, concentrating on these primes. K embeds into $K_{p^*} =_{def} Ult_U(\overline{K}_p)$, and K_{p^*} has an order l-point c_{p^*}. This point is rational over $\tilde{K}$, but not over $K\tilde{\mathbb{Q}}$. It follows that $K\tilde{\mathbb{Q}}$ is not relatively algebraically closed in $K_{p^*}\tilde{\mathbb{Q}}$. But this contradicts Lemma 3.

(One can also argue classically, as in [Lan 83]).

4) For almost all p, r induces an injective map on $A(K)/(lA(K))$.

This can be argued as in [Lan 83], or as follows. Suppose not, and find a non-principal ultrafilter U on the primes of L, such that (with the notation above) the map

$$A(Ult_U(K)/lA(Ult_U(K))) \to A(\overline{K}_{p^*})/lA(\overline{K}_{p^*})$$

is not injective.

On the other hand, by Lemma 3, $\tilde{K}$ is linearly disjoint from $\tilde{\mathbb{Q}}\overline{K}_{p^*}$ over $K\tilde{\mathbb{Q}}$. In particular, if an element of $A(K)$ has an l'th root in $A(\overline{K}_{p^*})$, it already has one in $A(K\tilde{\mathbb{Q}}) = A(K)$. Thus

$$A(K)/lA(K) \to A(\overline{K}_{p^*})/lA(\overline{K}_{p^*})$$

is injective.

But by Los's theorem, $A(K)/lA(K)$ maps bijectively into

$$A(Ult_U(K)/lA(Ult_U(K))).$$

Comparing these maps, we obtain a contradiction.

5) By Lemma 1, r is injective on any torsion-free part of $A(K)$. In other words, the kernel of r is contained in the torsion part of $A(K)$. Hence by 2), the kernel is trivial. This finishes the proof. □

Isotriviality The following lemma and its corollaries address the isotriviality issue. In positive characteristic, we will say that an abelian variety over k *has no weakly isotrivial factors* if there are no nonzero homomorphisms (defined over $\tilde{k}$) from an abelian variety defined over a finite field, to A. The word "weakly" refers to the fact that we do not insist that this homomorphism be separable.

Lemma 5 *Let A be a semi-abelian variety defined over a field F, B a commutative group variety defined over F, $h : A \to B$ a definable homomorphism, over some extension of F. Let $\{b_i\}$ be the set of points of $A(\tilde{F})$ or $B(\tilde{F})$ of order l or l', where l,l' are two distinct primes, and distinct from the characteristic. Let $b = (b_i : i)$. Then h is defined over $F(b)$.*

Proof: In any case h is defined over some finite extension of $F(b)$, and there is a unique minimal extension M over which h is defined ("Chow coordinates"). Using Weil's theorem on symmetric functions, one can show that M/F is separable (with a proof similar to Lang's proof of Chow's theorem, Chapter II, Theorem 5 in [Lan 59]).

Let $F*(l)$ be the maximal Galois extension of $F(b)$ whose Galois group is an l-group. Then all l^m-torsion points of A or B are rational over $F*(l)$. (Show this by induction on m. For $m = 1$, this is the definition of $F(b) \subset F*(l)$. Once we know the l^m-torsion points are rational over $F*(l)$, $m \geq 1$, let R be the field generated by the l^{m+1}-torsion points over $F*(l)$. For any automorphism σ of R over $F*(l)$, and any l^{m+1}-torsion point c, we have $\sigma(lc) = lc$, and so $\sigma(c) - c = \alpha$, α an l-torsion point, hence rational over $F*(l)$. It follows that the Galois group of R over $F*(l)$ is an l-group, a quotient of a power of the group of l-torsion points. But then by definition of $F*(l)$, we have $R = F*(l)$.) We will show that $M \subset F*(l)$. Similarly $M \subset F*(l')$; since $F*(l)$ and $F*(l')$ are linearly disjoint over $F(b)$, $M = F(b)$. We must show that h has no proper conjugates over $F*(l)$. Suppose for contradiction that h' is such a conjugate. Then $E = Ker(h-h')$ is a proper subgroup of A. It follows that there are points a in $A - E$ of order a power of l. (Let E^0 be the connected component of E. There exist points in A/E^0, of order any power of l; since E^0 is divisible, such points exist in A.) Thus $h(a)$ and $h'(a)$ are torsion points of B of order a power of l, and so they are rational over $F*(l)$. However there exists an automorphism τ fixing $F*(l)$, with $\tau(h) = h'$. So $h(a) = \tau(h(a)) = h'(\tau(a)) = h'(a)$, so $a \in E$, a contradiction. □

Corollary 6 *Every homomorphism $h : A \to B$ is defined over a field extension of F of degree bounded by β, where β depends only on $dim(A)$ and $dim(B)$. If B is defined over an algebraically closed subfield of F, then β depends only on $dim(A)$.*

Corollary 7 *Let F_0 be an algebraically closed field, B a semi-abelian variety defined over F_0. Let F be a field extension of F_0, A a semi-abelian variety defined over F, $h : B \to A$ a definable homomorphism. Then all torsion points of $h(B)$ are contained in a finite extension of F of degree at most β, where $\beta = \beta(dim(A))$ depends only on $dim(A)$.*

Corollary 8 *Let A be a (semi) abelian variety defined over $L(c)$, where L is a number field, and c is a generic point of a variety V defined over $\mathcal{O}_L$. Then for almost all primes p of $\mathcal{O}_L$, if $V(p)$ is the reduction of V modulo p, and $c(p)$ is a generic point of $V(p)$, and $A(p)$ is the reduced variety, then $A(p)$ is a (semi) abelian variety. If A is an abelian variety with no isotrivial components, then $A(p)$ has no (weakly) isotrivial components.*

Proof: By Lemma 2 $A(p)$ is an irreducible group variety, of dimension $d = dim(A)$, for almost all p. The fact that it remains an abelian, or semi-abelian, variety can be ascertained by counting 2-torsion points. Alternatively, the toric part of $A(p)$ remains isomorphic to a torus (after a finite field extension) using the Lefschetz principle, while as noted earlier, the quotient abelian variety remains projective and hence an abelian variety.

Now suppose A is an abelian variety with no isotrivial components. Let $\beta = \beta(dim(A))$ be as in the previous corollary. Let $k = \mathcal{O}_L/p$. Let A_m be the set of elements of A of order exactly m, i.e. $na = 0$ iff $m|n$. Suppose there exist $e_m \in A_m$ with $[\tilde{L}(c, e_m) : \tilde{L}(c)] \leq \beta]$; by König's lemma, there exists an infinite set of integers S (powers of a fixed prime) such that e_m is a multiple of e_l when $m < l \in S$. But then $\tilde{L}(c)(e_m) \subset \tilde{L}(c)(e_l)$ when $m < l \in S$, so the tower of fields stabilizes at some finite stage, and hence infinitely many torsion points of A lie in a finite extension of $\tilde{L}(c)$, contradicting the Lang-Neron Mordel-Weil theorem for that finite extension.

Thus there exists m such that:

(*) For any $e \in A_m$, $[\tilde{L}(c, e) : \tilde{L}(c)] > \beta$.

For almost all p, the same statement (*) holds mod p, i.e. for any $\overline{e} \in A_m(p)$

$$[\tilde{k}(c(p), \overline{e}) : \tilde{k}(c(p))] > \beta$$

Now suppose $A(p)$ has a weakly isotrivial component. Then there exists a nonzero definable homomorphism $h : B \to A(p)$ with B an abelian variety defined over a finite field. By the previous lemma, h can be defined over a finite extension of $\tilde{k}(c(p))$ of degree bounded by β. But $h(B)$ contains torsion points of all orders, in particular one of order m. This contradiction shows that $A(p)$ has no weakly isotrivial component. □

Uniform definition of subgroups Let ACF be the theory of algebraically closed fields. ACF is incomplete, and the completions correspond to the possible characteristics of fields. If Morley rank for an incomplete theory is reasonably defined, ACF has Morley rank two. It follows in particular that ACF is stable; for each formula $\phi(x, y)$, for some m, ACF proves that ϕ does not linearly order a set of size m. Note that this is equivalent to the stability of each completion.

Recall Shelah's property "nfcp" (negation of the finite cover property). This property states that for any formula $\phi(x, y)$ there exists a bound $b = b(\phi)$ with the following property. Let C be a collection of definable sets of the form $\{y : \phi(a, y)\}$. If every subset of C of size $\leq b$ has nonempty intersection, then so does C. Unlike stability, this property does not follow from its truth in each completion. However, it is true for ACF, or for any incomplete theory all of whose completions have Morley rank one.

Lemma 9 *ACF has the property nfcp*

Proof: A weaker form of nfcp, wnfcp, states that for each $\phi(x, y)$ there exists b' such that if $\phi(a, y)$ has finitely many solutions, then it has at most b' solutions. The proofs in [BaLa] or in [Hr 89] that non-multi-dimensional stable theories have the wnfcp go through for incomplete theories. So does Shelah's proof [She] that for stable theories, wnfcp implies nfcp. □

The following remark, like many other improvements throughout the paper, is due to Zoe Chatzidakis and Dave Marker. It shows that Lemma 9 can be dispensed with in the present context.

Remark 10 *Consider an algebraic family of* closed *subsets C_a of a variety V over an algebraically closed field K. There exists an integer k such that the intersection of any $k+1$ sets C_a is an intersection of at most k of them. This is an elementary property and hence transfers for almost all primes.*

Proof: The existence of k follows from Noetherianity and compactness; unlike the general "nfcp", it is clearly a first-order statement about the family. □

We will say that $\phi(a)$ ensures a statement about a to mean: whenever $\phi(a)$ holds in some model of ACF, so does the given statement.

Lemma 11 *Let X be a closed subset of the definable abelian group A over an algebraically closed field. Let $MC = MC(A, X)$ be the family of cosets of connected definable subgroups of A that are contained in X, and maximal, in the sense that no coset of a bigger connected group is contained in X. Let $MB = MB(A, X)$ be the family of subgroups B of A such that some coset of B is in MC. Then the elements of MB are uniformly definable.*

If A is semi-abelian, then $MB = \{B_1, .., B_k\}$ is finite. Moreover, there is a first order formula true of the defining parameters of A,X, that ensures that $MB = \{B_1, .., B_k\}$.

Proof: If $B \in MB$, $B \neq 0$, consider $X' = \{x \in X : B + x \subset X\}$. X' is an intersection of sets $X - b$, hence a finite such intersection; in particular it is closed. If $x \in X'$, then $B + B + x \subset X$, so $B + x \subset X'$; thus $B \subset \text{Stab}(X')$. For some $x \in X', B + x \in MC$; we have $B + x \subset \text{Stab}(X') + x \subset X$, so by maximality $B = \text{Stab}(X')$. So B is the intersection of all sets $X' - x$, $x \in X'$. So B is itself an intersection of sets of the form $X + c$. By 10, B is the intersection of a bounded number of sets $X + c$, and this bound holds good as A,X vary through a uniformly definable family of definable groups and subsets, in models of ACF. Thus the elements of the set $MB(A, X)$ are uniformly definable in this situation. To be precise, there exists a uniformly definable family of subsets of A containing $MB(A, X)$, namely the intersections of the translates of X; among these the elements of $MB(A, X)$ can be picked out by a first order formula, stating that they are subgroups, having a coset contained in X, and maximal among cosets of members of $MB(A, X)$ contained in X.

If A is semi-abelian, then there is no infinite, uniformly definable family of subgroups of A (see Hindry's contribution to this volume for the abelian case; the proof using torsion points goes through for semi-abelian varieties). Thus $MB(A, X)$ is finite. The first order statement "every element of MB(A,X) is one of the subgroups B_i" is true of the parameters defining A, hence can be ensured by some sentence. □

Proof of the theorem We have an abelian variety A over K; a subvariety X; let $\Gamma = A(K)$. For all but finitely many primes p of L, we also have a reduction homomorphism $r : A(K) \to \overline{A}(\overline{K})$ where $\overline{A}$ is the reduction of A; the dependence on p will not be denoted explicitly.

Let $MB(A,X) = \{B_1, .., B_k\}$ be the set of definable subgroups of A mentioned in 11. For all but finitely many p, $MB(\overline{A},\overline{X}) = \{\overline{B_1}, \ldots, \overline{B_k}\}$. Let

$$S_i = \{b + B_i \in (A/B_i) : (b + B_i) \subset X\}$$

For each i, for sufficiently large p,

$$\overline{S_i} = \{b + \overline{B_i} \in (\overline{A}/\overline{B_i}) : (b + \overline{B_i}) \subset \overline{X}\}$$

By 4 (applied to each quotient A/B_i), for all large enough p, the reduction homomorphism $h_i : (A/B_i) \to (\overline{A}/\overline{B_i})$ can be chosen injective on the image of Γ mod B_i.

Choose p large enough for these facts to hold, and also such that $\overline{A}$ has no weakly isotrivial factors (8). By the characteristic p Mordell-Lang theorem, [Hr 96],

$$r(\Gamma) \cap \overline{X} \subset \cup_{i=1}^{M} C_j$$

where the C_j are cosets of definable subgroups of $\overline{A}$, contained in $\overline{X}$. Enlarging C_j within $\overline{X}$, we can assume each $C_j \in MC(\overline{A},\overline{X})$, so C_j is a coset of some $\overline{B_i} \in MB(\overline{A},\overline{X})$.

Let $p_i : A \to A/B_i$ be the canonical homomorphism; then we have finite subsets F_i of $\overline{S_i}$ such that for every element a of $r(\Gamma) \cap \overline{X}$, for some i, $p_i(a) \in F_i$.

We may assume $F_i \subset p_i(r(\Gamma))$, and lift F_i to a finite subset of $p_i(\Gamma)$. If there exists $a \in (\Gamma \cap X)$ such that $p_i(a) \notin F_i$ for any i, then by the injectivity of h_i on $p_i(\Gamma)$, $h_i(p_i(a)) \notin h_i(F_i)$; a contradiction. Thus for each $a \in (\Gamma \cap X)$, for some i, $p_i(a) \in F_i$; in other words a lies in a coset of B_i contained in X; and the number of such cosets is at most M. This finishes the proof. □

References

[BaLa] J. Baldwin and A. Lachlan, *On Strongly Minimal Sets*, J. Symbolic Logic 36 (1971), 70-96.

[vdDS] L. van den Dries and K. Schmidt, *Bounds in the theory of polynomial rings over fields, a non-standard approach*, Invent. Math. 76 (1984), 77-91.

[Ha] R. Hartshorne, *Algebraic Geometry*, Springer, 1977.

[Hr 89] E. Hrushovski, *Kueker's conjecture for stable theories*, J. Symbolic Logic 54 (1989), 221-225.

[Hr 96] E. Hrushovski, *The Mordell-Lang conjecture for function fields*, J. AMS 9 (1996), 667-690.

[Lan 59] S. Lang, *Abelian varieties*, Interscience, New York 1959.

[Lan 83] S. Lang, *Fundamentals of Diophantine Geometry*, Springer, 1983.

[Lan 91] S. Lang, *Number Theory III: Diophantine Geometry*, Encyclopedia of Mathematical Sciences, Springer, 1991.

[Man] Y. Manin, *Rational points of algebraic curves over function fields,* Isvetzia 27(1963), 1395-1440 (AMS Transl. Ser II 50 (1966) 189-234).

[Noe] E. Noether, *Eliminationstheorie und allgemeine Idealtheorie*, Math. Annalen (1923), 229-261.

[She] S. Shelah, *Classification Theory*, revised edition, Studies in Logic 92, North-Holland, 1990.

Index

Lecture Notes in Mathematics

For information about Vols. 1–1504
please contact your bookseller or Springer-Verlag

Vol. 1505: K. Kajitani, T. Nishitani, The Hyperbolic Cauchy Problem. VII, 168 pages. 1991.

Vol. 1506: A. Buium, Differential Algebraic Groups of Finite Dimension. XV, 145 pages. 1992.

Vol. 1507: K. Hulek, T. Peternell, M. Schneider, F.-O. Schreyer (Eds.), Complex Algebraic Varieties. Proceedings, 1990. VII, 179 pages. 1992.

Vol. 1508: M. Vuorinen (Ed.), Quasiconformal Space Mappings. A Collection of Surveys 1960-1990. IX, 148 pages. 1992.

Vol. 1509: J. Aguadé, M. Castellet, F. R. Cohen (Eds.), Algebraic Topology - Homotopy and Group Cohomology. Proceedings, 1990. X, 330 pages. 1992.

Vol. 1510: P. P. Kulish (Ed.), Quantum Groups. Proceedings, 1990. XII, 398 pages. 1992.

Vol. 1511: B. S. Yadav, D. Singh (Eds.), Functional Analysis and Operator Theory. Proceedings, 1990. VIII, 223 pages. 1992.

Vol. 1512: L. M. Adleman, M.-D. A. Huang, Primality Testing and Abelian Varieties Over Finite Fields. VII, 142 pages. 1992.

Vol. 1513: L. S. Block, W. A. Coppel, Dynamics in One Dimension. VIII, 249 pages. 1992.

Vol. 1514: U. Krengel, K. Richter, V. Warstat (Eds.), Ergodic Theory and Related Topics III, Proceedings, 1990. VIII, 236 pages. 1992.

Vol. 1515: E. Ballico, F. Catanese, C. Ciliberto (Eds.), Classification of Irregular Varieties. Proceedings, 1990. VII, 149 pages. 1992.

Vol. 1516: R. A. Lorentz, Multivariate Birkhoff Interpolation. IX, 192 pages. 1992.

Vol. 1517: K. Keimel, W. Roth, Ordered Cones and Approximation. VI, 134 pages. 1992.

Vol. 1518: H. Stichtenoth, M. A. Tsfasman (Eds.), Coding Theory and Algebraic Geometry. Proceedings, 1991. VIII, 223 pages. 1992.

Vol. 1519: M. W. Short, The Primitive Soluble Permutation Groups of Degree less than 256. IX, 145 pages. 1992.

Vol. 1520: Yu. G. Borisovich, Yu. E. Gliklikh (Eds.), Global Analysis – Studies and Applications V. VII, 284 pages. 1992.

Vol. 1521: S. Busenberg, B. Forte, H. K. Kuiken, Mathematical Modelling of Industrial Process. Bari, 1990. Editors: V. Capasso, A. Fasano. VII, 162 pages. 1992.

Vol. 1522: J.-M. Delort, F. B. I. Transformation. VII, 101 pages. 1992.

Vol. 1523: W. Xue, Rings with Morita Duality. X, 168 pages. 1992.

Vol. 1524: M. Coste, L. Mahé, M.-F. Roy (Eds.), Real Algebraic Geometry. Proceedings, 1991. VIII, 418 pages. 1992.

Vol. 1525: C. Casacuberta, M. Castellet (Eds.), Mathematical Research Today and Tomorrow. VII, 112 pages. 1992.

Vol. 1526: J. Azéma, P. A. Meyer, M. Yor (Eds.), Séminaire de Probabilités XXVI. X, 633 pages. 1992.

Vol. 1527: M. I. Freidlin, J.-F. Le Gall, Ecole d'Eté de Probabilités de Saint-Flour XX – 1990. Editor: P. L. Hennequin. VIII, 244 pages. 1992.

Vol. 1528: G. Isac, Complementarity Problems. VI, 297 pages. 1992.

Vol. 1529: J. van Neerven, The Adjoint of a Semigroup of Linear Operators. X, 195 pages. 1992.

Vol. 1530: J. G. Heywood, K. Masuda, R. Rautmann, S. A. Solonnikov (Eds.), The Navier-Stokes Equations II – Theory and Numerical Methods. IX, 322 pages. 1992.

Vol. 1531: M. Stoer, Design of Survivable Networks. IV, 206 pages. 1992.

Vol. 1532: J. F. Colombeau, Multiplication of Distributions. X, 184 pages. 1992.

Vol. 1533: P. Jipsen, H. Rose, Varieties of Lattices. X, 162 pages. 1992.

Vol. 1534: C. Greither, Cyclic Galois Extensions of Commutative Rings. X, 145 pages. 1992.

Vol. 1535: A. B. Evans, Orthomorphism Graphs of Groups. VIII, 114 pages. 1992.

Vol. 1536: M. K. Kwong, A. Zettl, Norm Inequalities for Derivatives and Differences. VII, 150 pages. 1992.

Vol. 1537: P. Fitzpatrick, M. Martelli, J. Mawhin, R. Nussbaum, Topological Methods for Ordinary Differential Equations. Montecatini Terme, 1991. Editors: M. Furi, P. Zecca. VII, 218 pages. 1993.

Vol. 1538: P.-A. Meyer, Quantum Probability for Probabilists. X, 287 pages. 1993.

Vol. 1539: M. Coornaert, A. Papadopoulos, Symbolic Dynamics and Hyperbolic Groups. VIII, 138 pages. 1993.

Vol. 1540: H. Komatsu (Ed.), Functional Analysis and Related Topics, 1991. Proceedings. XXI, 413 pages. 1993.

Vol. 1541: D. A. Dawson, B. Maisonneuve, J. Spencer, Ecole d´ Eté de Probabilités de Saint-Flour XXI - 1991. Editor: P. L. Hennequin. VIII, 356 pages. 1993.

Vol. 1542: J.Fröhlich, Th.Kerler, Quantum Groups, Quantum Categories and Quantum Field Theory. VII, 431 pages. 1993.

Vol. 1543: A. L. Dontchev, T. Zolezzi, Well-Posed Optimization Problems. XII, 421 pages. 1993.

Vol. 1544: M.Schürmann, White Noise on Bialgebras. VII, 146 pages. 1993.

Vol. 1545: J. Morgan, K. O'Grady, Differential Topology of Complex Surfaces. VIII, 224 pages. 1993.

Vol. 1546: V. V. Kalashnikov, V. M. Zolotarev (Eds.), Stability Problems for Stochastic Models. Proceedings, 1991. VIII, 229 pages. 1993.

Vol. 1547: P. Harmand, D. Werner, W. Werner, M-ideals in Banach Spaces and Banach Algebras. VIII, 387 pages. 1993.

Vol. 1548: T. Urabe, Dynkin Graphs and Quadrilateral Singularities. VI, 233 pages. 1993.

Vol. 1549: G. Vainikko, Multidimensional Weakly Singular Integral Equations. XI, 159 pages. 1993.

Vol. 1550: A. A. Gonchar, E. B. Saff (Eds.), Methods of Approximation Theory in Complex Analysis and Mathematical Physics IV, 222 pages, 1993.

Vol. 1551: L. Arkeryd, P. L. Lions, P.A. Markowich, S.R. S. Varadhan. Nonequilibrium Problems in Many-Particle Systems. Montecatini, 1992. Editors: C. Cercignani, M. Pulvirenti. VII, 158 pages 1993.

Vol. 1552: J. Hilgert, K.-H. Neeb, Lie Semigroups and their Applications. XII, 315 pages. 1993.

Vol. 1553: J.-L- Colliot-Thélène, J. Kato, P. Vojta. Arithmetic Algebraic Geometry. Trento, 1991. Editor: E. Ballico. VII, 223 pages. 1993.

Vol. 1554: A. K. Lenstra, H. W. Lenstra, Jr. (Eds.), The Development of the Number Field Sieve. VIII, 131 pages. 1993.

Vol. 1555: O. Liess, Conical Refraction and Higher Microlocalization. X, 389 pages. 1993.

Vol. 1556: S. B. Kuksin, Nearly Integrable Infinite-Dimensional Hamiltonian Systems. XXVII, 101 pages. 1993.

Vol. 1557: J. Azéma, P. A. Meyer, M. Yor (Eds.), Séminaire de Probabilités XXVII. VI, 327 pages. 1993.

Vol. 1558: T. J. Bridges, J. E. Furter, Singularity Theory and Equivariant Symplectic Maps. VI, 226 pages. 1993.

Vol. 1559: V. G. Sprindžuk, Classical Diophantine Equations. XII, 228 pages. 1993.

Vol. 1560: T. Bartsch, Topological Methods for Variational Problems with Symmetries. X, 152 pages. 1993.

Vol. 1561: I. S. Molchanov, Limit Theorems for Unions of Random Closed Sets. X, 157 pages. 1993.

Vol. 1562: G. Harder, Eisensteinkohomologie und die Konstruktion gemischter Motive. XX, 184 pages. 1993.

Vol. 1563: E. Fabes, M. Fukushima, L. Gross, C. Kenig, M. Röckner, D. W. Stroock, Dirichlet Forms. Varenna, 1992. Editors: G. Dell'Antonio, U. Mosco. VII, 245 pages. 1993.

Vol. 1564: J. Jorgenson, S. Lang, Basic Analysis of Regularized Series and Products. IX, 122 pages. 1993.

Vol. 1565: L. Boutet de Monvel, C. De Concini, C. Procesi, P. Schapira, M. Vergne. D-modules, Representation Theory, and Quantum Groups. Venezia, 1992. Editors: G. Zampieri, A. D'Agnolo. VII, 217 pages. 1993.

Vol. 1566: B. Edixhoven, J.-H. Evertse (Eds.), Diophantine Approximation and Abelian Varieties. XIII, 127 pages. 1993.

Vol. 1567: R. L. Dobrushin, S. Kusuoka, Statistical Mechanics and Fractals. VII, 98 pages. 1993.

Vol. 1568: F. Weisz, Martingale Hardy Spaces and their Application in Fourier Analysis. VIII, 217 pages. 1994.

Vol. 1569: V. Totik, Weighted Approximation with Varying Weight. VI, 117 pages. 1994.

Vol. 1570: R. deLaubenfels, Existence Families, Functional Calculi and Evolution Equations. XV, 234 pages. 1994.

Vol. 1571: S. Yu. Pilyugin, The Space of Dynamical Systems with the C^0-Topology. X, 188 pages. 1994.

Vol. 1572: L. Göttsche, Hilbert Schemes of Zero-Dimensional Subschemes of Smooth Varieties. IX, 196 pages. 1994.

Vol. 1573: V. P. Havin, N. K. Nikolski (Eds.), Linear and Complex Analysis – Problem Book 3 – Part I. XXII, 489 pages. 1994.

Vol. 1574: V. P. Havin, N. K. Nikolski (Eds.), Linear and Complex Analysis – Problem Book 3 – Part II. XXII, 507 pages. 1994.

Vol. 1575: M. Mitrea, Clifford Wavelets, Singular Integrals, and Hardy Spaces. XI, 116 pages. 1994.

Vol. 1576: K. Kitahara, Spaces of Approximating Functions with Haar-Like Conditions. X, 110 pages. 1994.

Vol. 1577: N. Obata, White Noise Calculus and Fock Space. X, 183 pages. 1994.

Vol. 1578: J. Bernstein, V. Lunts, Equivariant Sheaves and Functors. V, 139 pages. 1994.

Vol. 1579: N. Kazamaki, Continuous Exponential Martingales and *BMO*. VII, 91 pages. 1994.

Vol. 1580: M. Milman, Extrapolation and Optimal Decompositions with Applications to Analysis. XI, 161 pages. 1994.

Vol. 1581: D. Bakry, R. D. Gill, S. A. Molchanov, Lectures on Probability Theory. Editor: P. Bernard. VIII, 420 pages. 1994.

Vol. 1582: W. Balser, From Divergent Power Series to Analytic Functions. X, 108 pages. 1994.

Vol. 1583: J. Azéma, P. A. Meyer, M. Yor (Eds.), Séminaire de Probabilités XXVIII. VI, 334 pages. 1994.

Vol. 1584: M. Brokate, N. Kenmochi, I. Müller, J. F. Rodriguez, C. Verdi, Phase Transitions and Hysteresis. Montecatini Terme, 1993. Editor: A. Visintin. VII, 291 pages. 1994.

Vol. 1585: G. Frey (Ed.), On Artin's Conjecture for Odd 2-dimensional Representations. VIII, 148 pages. 1994.

Vol. 1586: R. Nillsen, Difference Spaces and Invariant Linear Forms. XII, 186 pages. 1994.

Vol. 1587: N. Xi, Representations of Affine Hecke Algebras. VIII, 137 pages. 1994.

Vol. 1588: C. Scheiderer, Real and Étale Cohomology. XXIV, 273 pages. 1994.

Vol. 1589: J. Bellissard, M. Degli Esposti, G. Forni, S. Graffi, S. Isola, J. N. Mather, Transition to Chaos in Classical and Quantum Mechanics. Montecatini Terme, 1991. Editor: 2S. Graffi. VII, 192 pages. 1994.

Vol. 1590: P. M. Soardi, Potential Theory on Infinite Networks. VIII, 187 pages. 1994.

Vol. 1591: M. Abate, G. Patrizio, Finsler Metrics – A Global Approach. IX, 180 pages. 1994.

Vol. 1592: K. W. Breitung, Asymptotic Approximations for Probability Integrals. IX, 146 pages. 1994.

Vol. 1593: J. Jorgenson & S. Lang, D. Goldfeld, Explicit Formulas for Regularized Products and Series. VIII, 154 pages. 1994.

Vol. 1594: M. Green, J. Murre, C. Voisin, Algebraic Cycles and Hodge Theory. Torino, 1993. Editors: A. Albano, F. Bardelli. VII, 275 pages. 1994.

Vol. 1595: R.D.M. Accola, Topics in the Theory of Riemann Surfaces. IX, 105 pages. 1994.

Vol. 1596: L. Heindorf, L. B. Shapiro, Nearly Projective Boolean Algebras. X, 202 pages. 1994.

Vol. 1597: B. Herzog, Kodaira-Spencer Maps in Local Algebra. XVII, 176 pages. 1994.

Vol. 1598: J. Berndt, F. Tricerri, L. Vanhecke, Generalized Heisenberg Groups and Damek-Ricci Harmonic Spaces. VIII, 125 pages. 1995.

Vol. 1599: K. Johannson, Topology and Combinatorics of 3-Manifolds. XVIII, 446 pages. 1995.

Vol. 1600: W. Narkiewicz, Polynomial Mappings. VII, 130 pages. 1995.

Vol. 1601: A. Pott, Finite Geometry and Character Theory. VII, 181 pages. 1995.

Vol. 1602: J. Winkelmann, The Classification of Three-dimensional Homogeneous Complex Manifolds. XI, 230 pages. 1995.

Vol. 1603: V. Ene, Real Functions – Current Topics. XIII, 310 pages. 1995.

Vol. 1604: A. Huber, Mixed Motives and their Realization in Derived Categories. XV, 207 pages. 1995.

Vol. 1605: L. B. Wahlbin, Superconvergence in Galerkin Finite Element Methods. XI, 166 pages. 1995.

Vol. 1606: P.-D. Liu, M. Qian, Smooth Ergodic Theory of Random Dynamical Systems. XI, 221 pages. 1995.

Vol. 1607: G. Schwarz, Hodge Decomposition – A Method for Solving Boundary Value Problems. VII, 155 pages. 1995.

Vol. 1608: P. Biane, R. Durrett, Lectures on Probability Theory. Editor: P. Bernard. VII, 210 pages. 1995.

Vol. 1609: L. Arnold, C. Jones, K. Mischaikow, G. Raugel, Dynamical Systems. Montecatini Terme, 1994. Editor: R. Johnson. VIII, 329 pages. 1995.

Vol. 1610: A. S. Üstünel, An Introduction to Analysis on Wiener Space. X, 95 pages. 1995.

Vol. 1611: N. Knarr, Translation Planes. VI, 112 pages. 1995.

Vol. 1612: W. Kühnel, Tight Polyhedral Submanifolds and Tight Triangulations. VII, 122 pages. 1995.

Vol. 1613: J. Azéma, M. Emery, P. A. Meyer, M. Yor (Eds.), Séminaire de Probabilités XXIX. VI, 326 pages. 1995.

Vol. 1614: A. Koshelev, Regularity Problem for Quasilinear Elliptic and Parabolic Systems. XXI, 255 pages. 1995.

Vol. 1615: D. B. Massey, Le Cycles and Hypersurface Singularities. XI, 131 pages. 1995.

Vol. 1616: I. Moerdijk, Classifying Spaces and Classifying Topoi. VII, 94 pages. 1995.

Vol. 1617: V. Yurinsky, Sums and Gaussian Vectors. XI, 305 pages. 1995.

Vol. 1618: G. Pisier, Similarity Problems and Completely Bounded Maps. VII, 156 pages. 1996.

Vol. 1619: E. Landvogt, A Compactification of the Bruhat-Tits Building. VII, 152 pages. 1996.

Vol. 1620: R. Donagi, B. Dubrovin, E. Frenkel, E. Previato, Integrable Systems and Quantum Groups. Montecatini Terme, 1993. Editors:M. Francaviglia, S. Greco. VIII, 488 pages. 1996.

Vol. 1621: H. Bass, M. V. Otero-Espinar, D. N. Rockmore, C. P. L. Tresser, Cyclic Renormalization and Auto-morphism Groups of Rooted Trees. XXI, 136 pages. 1996.

Vol. 1622: E. D. Farjoun, Cellular Spaces, Null Spaces and Homotopy Localization. XIV, 199 pages. 1996.

Vol. 1623: H.P. Yap, Total Colourings of Graphs. VIII, 131 pages. 1996.

Vol. 1624: V. Brinzanescu, Holomorphic Vector Bundles over Compact Complex Surfaces. X, 170 pages. 1996.

Vol.1625: S. Lang, Topics in Cohomology of Groups. VII, 226 pages. 1996.

Vol. 1626: J. Azéma, M. Emery, M. Yor (Eds.), Séminaire de Probabilités XXX. VIII, 382 pages. 1996.

Vol. 1627: C. Graham, Th. G. Kurtz, S. Méléard, Ph. E. Protter, M. Pulvirenti, D. Talay, Probabilistic Models for Nonlinear Partial Differential Equations. Montecatini Terme, 1995. Editors: D. Talay, L. Tubaro. X, 301 pages. 1996.

Vol. 1628: P.-H. Zieschang, An Algebraic Approach to Association Schemes. XII, 189 pages. 1996.

Vol. 1629: J. D. Moore, Lectures on Seiberg-Witten Invariants. VII, 105 pages. 1996.

Vol. 1630: D. Neuenschwander, Probabilities on the Heisenberg Group: Limit Theorems and Brownian Motion. VIII, 139 pages. 1996.

Vol. 1631: K. Nishioka, Mahler Functions and Transcendence. VIII, 185 pages. 1996.

Vol. 1632: A. Kushkuley, Z. Balanov, Geometric Methods in Degree Theory for Equivariant Maps. VII, 136 pages. 1996.

Vol.1633: H. Aikawa, M. Essén, Potential Theory – Selected Topics. IX, 200 pages. 1996.

Vol. 1634: J. Xu, Flat Covers of Modules. IX, 161 pages. 1996.

Vol. 1635: E. Hebey, Sobolev Spaces on Riemannian Manifolds. X, 116 pages. 1996.

Vol. 1636: M. A. Marshall, Spaces of Orderings and Abstract Real Spectra. VI, 190 pages. 1996.

Vol. 1637: B. Hunt, The Geometry of some special Arithmetic Quotients. XIII, 332 pages. 1996.

Vol. 1638: P. Vanhaecke, Integrable Systems in the realm of Algebraic Geometry. VIII, 218 pages. 1996.

Vol. 1639: K. Dekimpe, Almost-Bieberbach Groups: Affine and Polynomial Structures. X, 259 pages. 1996.

Vol. 1640: G. Boillat, C. M. Dafermos, P. D. Lax, T. P. Liu, Recent Mathematical Methods in Nonlinear Wave Propagation. Montecatini Terme, 1994. Editor: T. Ruggeri. VII, 142 pages. 1996.

Vol. 1641: P. Abramenko, Twin Buildings and Applications to S-Arithmetic Groups. IX, 123 pages. 1996.

Vol. 1642: M. Puschnigg, Asymptotic Cyclic Cohomology. XXII, 138 pages. 1996.

Vol. 1643: J. Richter-Gebert, Realization Spaces of Polytopes. XI, 187 pages. 1996.

Vol. 1644: A. Adler, S. Ramanan, Moduli of Abelian Varieties. VI, 196 pages. 1996.

Vol. 1645: H. W. Broer, G. B. Huitema, M. B. Sevryuk, Quasi-Periodic Motions in Families of Dynamical Systems. XI, 195 pages. 1996.

Vol. 1646: J.-P. Demailly, T. Peternell, G. Tian, A. N. Tyurin, Transcendental Methods in Algebraic Geometry. Cetraro, 1994. Editors: F. Catanese, C. Ciliberto. VII, 257 pages. 1996.

Vol. 1647: D. Dias, P. Le Barz, Configuration Spaces over Hilbert Schemes and Applications. VII. 143 pages. 1996.

Vol. 1648: R. Dobrushin, P. Groeneboom, M. Ledoux, Lectures on Probability Theory and Statistics. Editor: P. Bernard. VIII, 300 pages. 1996.

Vol. 1649: S. Kumar, G. Laumon, U. Stuhler, Vector Bundles on Curves – New Directions. Cetraro, 1995. Editor: M. S. Narasimhan. VII, 193 pages. 1997.

Vol. 1650: J. Wildeshaus, Realizations of Polylogarithms. XI, 343 pages. 1997.

Vol. 1651: M. Drmota, R. F. Tichy, Sequences, Discrepancies and Applications. XIII, 503 pages. 1997.

Vol. 1652: S. Todorcevic, Topics in Topology. VIII, 153 pages. 1997.

Vol. 1653: R. Benedetti, C. Petronio, Branched Standard Spines of 3-manifolds. VIII, 132 pages. 1997.

Vol. 1654: R. W. Ghrist, P. J. Holmes, M. C. Sullivan, Knots and Links in Three-Dimensional Flows. X, 208 pages. 1997.

Vol. 1655: J. Azéma, M. Emery, M. Yor (Eds.), Séminaire de Probabilités XXXI. VIII, 329 pages. 1997.

Vol. 1656: B. Biais, T. Björk, J. Cvitanic, N. El Karoui, E. Jouini, J. C. Rochet, Financial Mathematics. Bressanone, 1996. Editor: W. J. Runggaldier. VII, 316 pages. 1997.

Vol. 1657: H. Reimann, The semi-simple zeta function of quaternionic Shimura varieties. IX, 143 pages. 1997.

Vol. 1658: A. Pumarino, J. A. Rodrıguez, Coexistence and Persistence of Strange Attractors. VIII, 195 pages. 1997.

Vol. 1659: V, Kozlov, V. Maz'ya, Theory of a Higher-Order Sturm-Liouville Equation. XI, 140 pages. 1997.

Vol. 1660: M. Bardi, M. G. Crandall, L. C. Evans, H. M. Soner, P. E. Souganidis, Viscosity Solutions and Applications. Montecatini Terme, 1995. Editors: I. Capuzzo Dolcetta, P. L. Lions. IX, 259 pages. 1997.

Vol. 1661: A. Tralle, J. Oprea, Symplectic Manifolds with no Kähler Structure. VIII, 207 pages. 1997.

Vol. 1662: J. W. Rutter, Spaces of Homotopy Self-Equivalences – A Survey. IX, 170 pages. 1997.

Vol. 1663: Y. E. Karpeshina; Perturbation Theory for the Schrödinger Operator with a Periodic Potential. VII, 352 pages. 1997.

Vol. 1664: M. Väth, Ideal Spaces. V, 146 pages. 1997.

Vol. 1665: E. Giné, G. R. Grimmett, L. Saloff-Coste, Lectures on Probability Theory and Statistics 1996. Editor: P. Bernard. X, 424 pages, 1997.

Vol. 1666: M. van der Put, M. F. Singer, Galois Theory of Difference Equations. VII, 179 pages. 1997.

Vol. 1667: J. M. F. Castillo, M. González, Three-space Problems in Banach Space Theory. XII, 267 pages. 1997.

Vol. 1668: D. B. Dix, Large-Time Behavior of Solutions of Linear Dispersive Equations. XIV, 203 pages. 1997.

Vol. 1669: U. Kaiser, Link Theory in Manifolds. XIV, 167 pages. 1997.

Vol. 1670: J. W. Neuberger, Sobolev Gradients and Differential Equations. VIII, 150 pages. 1997.

Vol. 1671: S. Bouc, Green Functors and G-sets. VII, 342 pages. 1997.

Vol. 1672: S. Mandal, Projective Modules and Complete Intersections. VIII, 114 pages. 1997.

Vol. 1673: F. D. Grosshans, Algebraic Homogeneous Spaces and Invariant Theory. VI, 148 pages. 1997.

Vol. 1674: G. Klaas, C. R. Leedham-Green, W. Plesken, Linear Pro-p-Groups of Finite Width. VIII, 115 pages. 1997.

Vol. 1675: J. E. Yukich, Probability Theory of Classical Euclidean Optimization Problems. X, 152 pages. 1998.

Vol. 1676: P. Cembranos, J. Mendoza, Banach Spaces of Vector-Valued Functions. VIII. 118 pages. 1997.

Vol. 1677: N. Proskurin, Cubic Metaplectic Forms and Theta Functions. VIII, 196 pages. 1998.

Vol. 1678: O. Krupková, The Geometry of Ordinary Variational Equations. X, 251 pages. 1997.

Vol. 1679: K.-G. Grosse-Erdmann, The Blocking Technique. Weighted Mean Operators and Hardy's Inequality. IX, 114 pages. 1998.

Vol. 1680: K.-Z. Li, F. Oort, Moduli of Supersingular Abelian Varieties. V, 116 pages. 1998.

Vol. 1681: G. J. Wirsching, The Dynamical System Generated by the 3n+1 Function. VII, 158 pages. 1998.

Vol. 1682: H.-D. Alber, Materials with Memory. X, 166 pages. 1998.

Vol. 1683: A. Pomp. The Boundary-Domain Integral Method for Elliptic Systems. XVI, 163 pages. 1998.

Vol. 1684: C. A. Berenstein, P. F. Ebenfelt, S. G. Gindikin, S. Helgason, A. E. Tumanov, Integral Geometry, Radon Transforms and Complex Analysis. Firenze, 1996. Editors: E. Casadio Tarabusi, M. A. Picardello, G. Zampieri. VII, 160 pages. 1998.

Vol. 1685: S. König, A. Zimmermann, Derived Equivalences for Group Rings. X, 146 pages. 1998.

Vol. 1686: J. Azéma, M. Émery, M. Ledoux, M. Yor (Eds.), Séminaire de Probabilités XXXII. VI, 440 pages. 1998.

Vol. 1687: F. Bornemann, Homogenization in Time of Singularly Perturbed Mechanical Systems. XII, 156 pages. 1998.

Vol. 1688: S. Assing. W. Schmidt. Continuous Strong Markov Processes in Dimension One. XII, 137 page. 1998.

Vol. 1689: W. Fulton, P. Pragacz, Schubert Varieties and Degeneracy Loci. XI, 148 pages. 1998.

Vol. 1690: M. T. Barlow, D. Nualart, Lectures on Probability Theory and Statistics. Editor: P. Bernard. VIII, 237 pages. 1998.

Vol. 1691: R. Bezrukavnikov, M. Finkelberg, V. Schechtman, Factorizable Sheaves and Quantum Groups. X, 282 pages. 1998.

Vol. 1692: T. M. W. Eyre, Quantum Stochastic Calculus and Representations of Lie Superalgebras. IX, 138 pages. 1998.

Vol. 1694: A. Braides, Approximation of Free-Discontinuity Problems. XI, 149 pages. 1998.

Vol. 1695: D. J. Hartfiel, Markov Set-Chains. VIII, 131 pages. 1998.

Vol. 1696: E. Bouscaren (Ed.): Model Theory and Algebraic Geometry. XV, 211 pages. 1998.